PARTIAL SOLUT

to accompany

ELEMENTARY STATISTICS

Fifth Edition

PWS-KENT
Publishing Company

20 Park Plaza
Boston, Massachusetts 02116

Copyright © 1988 by PWS-KENT Publishing Company.

All rights reserved. No part of this book may be reproduced or transmitted in any form or by any means, electronic or mechanical including photocopying, recording, or by any information storage and retrieval system, without permission, in writing, from the publisher.

PWS-KENT Publishing Company is a division of Wadsworth, Inc.

Printed in the United States of America.
90 91 92 – 10 9 8 7 6 5 4 3 2

ISBN 0-534-91773-9

PREFACE

The direction and the objectives of a first course in statistics can vary greatly. Consequently, there are various ways to solve most statistical problems. The solutions put forth in this manual represent only the author's point of view.

The purpose of this solutions manual is to provide the student with at least one complete solution for approximately every third exercise in <u>ELEMENTARY STATISTICS, Fifth Edition</u> by Robert Johnson {PWS-KENT Publishing Company, 1988}.

CONTENTS

Chapter 1 1
Chapter 2 3
Chapter 3 18
Chapter 4 23
Chapter 5 29
Chapter 6 38
Chapter 7 45
Chapter 8 49
Chapter 9 57
Chapter 10 65
Chapter 11 80
Chapter 12 85
Chapter 13 88
Chapter 14 98

Chapter One

1-1
(a) The amount of each type of throw away discarded by each person in the neighborhood.
(b) Weight of the throw aways, in pounds.
(c) It appears that they collected the amount of each type of throw away for each person for a given day.
(d) Weight measured in pounds.
(e) Percentage of total. Since pounds are mentioned, we would assume that the percentages are based on pounds.

1-4
(a) Those individuals who have a condition called hypertension. (A very large but unknown number.)
(b) The sample is the 5000 people in the study.
(c) The parameter is the proportion in the population for which the drug is effective.
(d) The statistic is the proportion in the sample for which the drug is effective. The eighty percent is the value of the statistic.
(e) The value of the parameter is unknown, but the 80% could be used to estimate its value.

1-6 (1) attribute   (2) continuous   (3) discrete

1-8
(a) Population - all students enrolled at your college this semester
(b) Variable - the amount of money each student spent to purchase their textbooks for this semester

1-9
(a) Population parameter - <u>average cost</u> of textbooks per student <u>for all students</u>
(b) Sample statistic - <u>average cost</u> of textbooks per student <u>for the 50</u> in the sample
(c) The <u>average cost</u> could be found by adding the cost of textbooks for all 50 students together and then dividing the sum by 50.

1-11 (a) continuous   (b) attribute   (c) discrete
(d) attribute   (e) discrete   (f) continuous

1-14 Group 2

1-16
(a) The sampling frame is that set of elements from which the sample is actually drawn.
(b) A computer list of this semester's full-time enrollment.

1-19 Stratified sampling - each supermarket chain is a strata and each strata is sampled.

1-21 (a) statistics   (b) probability
      (c) statistics   (d) probability

1-23 Several large computer programs (called statistical packages) have been developed which perform many of the statistical computations and tests which you will study in this text.
      In order to have the statistical package perform the computation or run the statistical test all you have to do is enter your data into the computer and it does the rest. This saves time and therefore money.

1-24 Each student's answers will differ. Some possibilities are:
      (a) color of hair, major, sex, marital status, hometown
      (b) number of courses currently enrolled in, number of semesters at college, number of roommates
      (c) height, weight, distance from hometown, cost of textbooks

1-27 Each of the numbers reported in A Liberty Blast are values representing the amount of tape, wire, etc. used to put on the fireworks display at the Statue of Liberty. The values reported do not result from repeated observations of variables.

1-30
(a) T = 3 is a data, the result obtained from one person.
(b) What is the average number of times that the people in the sample go grocery shopping per week?
(c) What is the average number of times that the people in the sampled population go grocery shopping per week?

1-32 Cluster sampling. The blocks are clusters or strata. Some, but not all of the clusters are selected to be part of the sample. In this example, cluster sampling would save on expense (time and travel) as compared with simple random sampling.

Chapter Two

2-1 "I enjoy using computers"

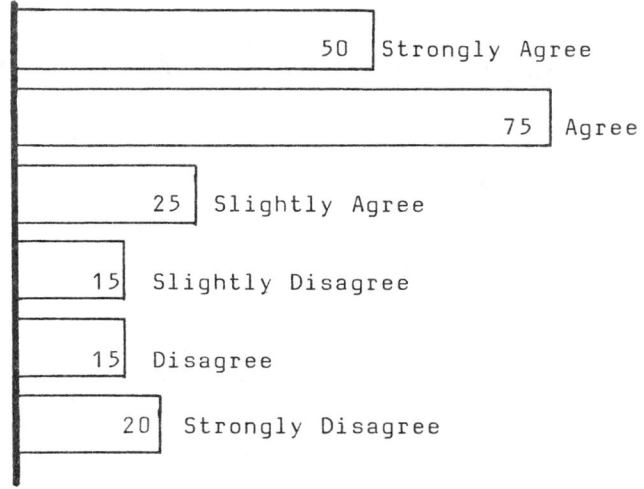

- 50 Strongly Agree
- 75 Agree
- 25 Slightly Agree
- 15 Slightly Disagree
- 15 Disagree
- 20 Strongly Disagree

2-4 (a) 15 patients (b) 11 days (c) 50 days (d) 20 days

2-6 Number of Detectable Emissions in 10 Seconds

```
0 | 8
1 | 8 5 8 9 2 6
2 | 3 2 2 1 3 5 4 2 1 2 2 7 6
3 | 7 2
```

2-10
(a,b,c)

Age	tallies	freq	rel freq	cum rel
17	I	1	0.02	0.02
18	III	3	0.06	0.08
19	IIII IIII IIII I	16	0.32	0.40
20	IIII IIII	10	0.20	0.60
21	IIII IIII II	12	0.24	0.84
22	IIII	5	0.10	0.94
23	I	1	0.02	0.96
24	II	2	0.04	1.00
		50	1.00	

(d)

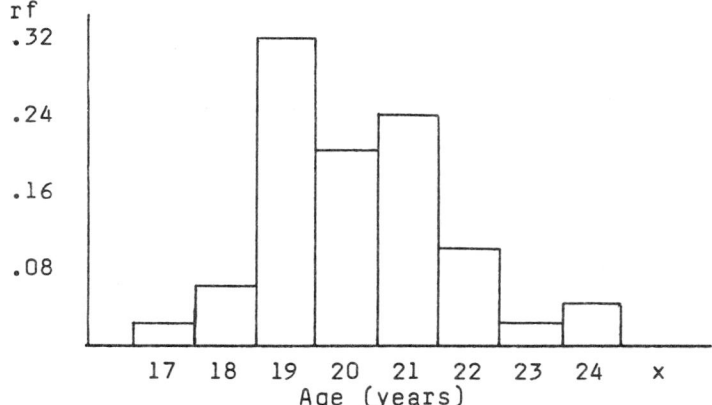

Ages of Dancers at Audition

(e)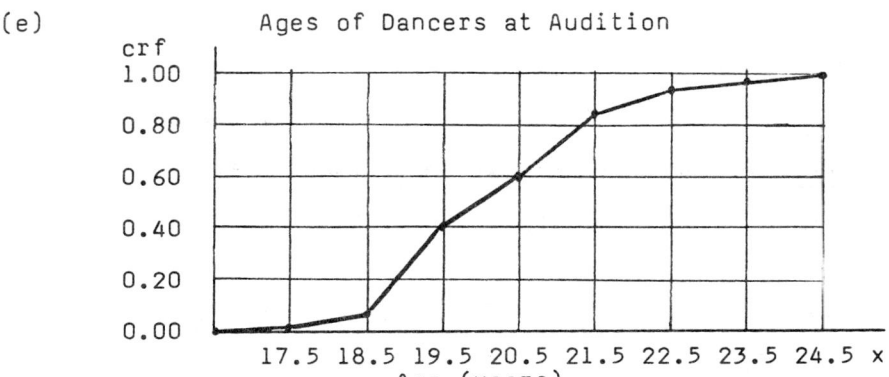

2-13
(a)

Class limits	tallies	f
15 - 19	IIII	4
20 - 24	LHT LHT IIII	14
25 - 29	LHT LHT LHT IIII	19
30 - 34	LHT II	7
35 - 39	LHT	5
40 - 44	III	3
45 - 49	II	2
50 - 54	I	1
		55

(b) Class width = 5

(c) Class mark = (20 + 24)/2 = 22
Lower class limit = 20
Upper class boundary = 24.5

(d)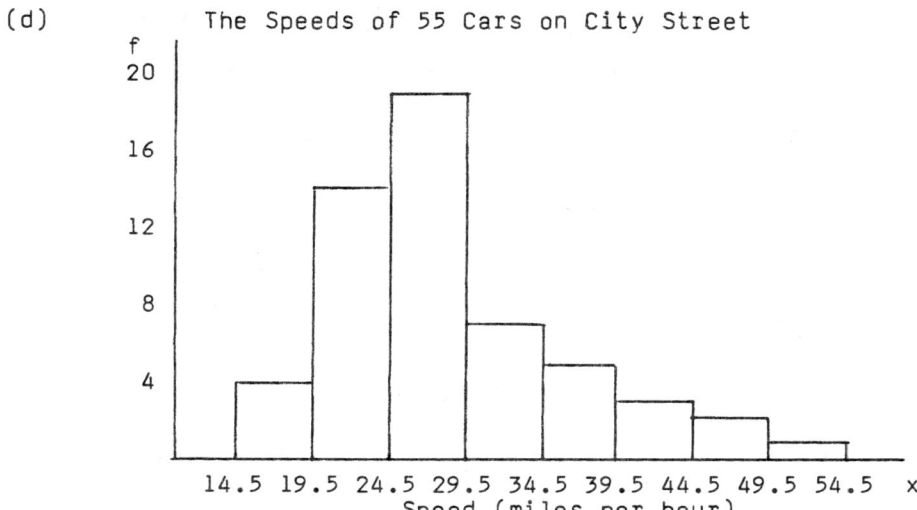

2-16   {2, 4, 7, 8, 9}

(a) $\bar{x} = \Sigma x/n = (2+4+7+8+9)/5 = 30/5 = \underline{6.0}$

(b) $\tilde{x} = \underline{7}$ ($i = (n+1)/2 = 3$, median is 3rd data)

(c)   <u>No mode</u>, no value appears more than once.

(d)   midrange = $(L+H)/2 = (2+9)/2 = \underline{5.5}$

2-19   {4, 5, 5, 6, 6, 6, 7, 7, 7, 7, 8, 8, 8, 9, 11}

(a) $\bar{x} = \Sigma x/n = (104)/15 = 6.9333 = \underline{6.9}$

(b) $\tilde{x} = \underline{7}$ (i = 8th)

(c)   mode = $\underline{7}$

(d)   midrange = $(4+11)/2 = \underline{7.5}$

2-21   ranked data: 25,500   31,500   31,500   31,500   31,500
                      35,250   36,750   37,500   39,000   54,000

Note: There is a typographical error in the data as listed in article. See the paragraph, the midrange, and specific reference is made to the largest data.

$\bar{x} = \Sigma x/n = 354,000/10 = \underline{35,400.00}$

$\tilde{x} = (31,500+35,250)/2 = \underline{33,375}$ ($i = (n+1)/2$ = $(10+1)/2 = 5.5$th, median is halfway between 5th and 6th data)

mode = $\underline{31,500}$

midrange = $(L+H)/2 = (25,500+54,000)/2 = \underline{39,750}$

The values of these statistics all agree.

2-23
(a)

x	f	xf
12.5	2	25.0
12.7	6	76.2
13.0	22	286.0
13.1	29	379.9
13.2	12	158.4
13.8	4	55.2
	75	980.7

$\bar{x} = \Sigma xf/n$

$\bar{x} = 980.7/75$

$= 13.076 = \underline{13.1}$

(b) $i = (n+1)/2 = (75+1)/2 = 38$th. $\tilde{x} = \underline{13.1}$

(c)   mode = $\underline{13.1}$

(d)   midrange = $(L+H)/2 = (12.5+13.8)/2 = \underline{13.15}$

2-26

Class limits	x	f	xf
15 - 19	17	4	68
20 - 24	22	14	308
25 - 29	27	19	513
30 - 34	32	7	224
35 - 39	37	5	185
40 - 44	42	3	126
45 - 49	47	2	94
50 - 54	52	1	52
sum		55	1570

$\bar{x} = \Sigma xf/n = 1570/55 = 28.545 = \underline{28.5}$

2-29   {2, 4, 7, 8, 9} $\bar{x} = 6.0$ from exercise 2-16.

(a)   range = 9 - 2 = $\underline{7}$

(b)

x	x-$\bar{x}$	(x-$\bar{x}$)²
2	-4	16
4	-2	4
7	1	1
8	2	4
9	3	9
sum	0	34

$s^2 = 34/4 = \underline{8.5}$

(c) $s = \sqrt{8.5} = 2.915 = \underline{2.9}$

2-32   {4, 5, 5, 6, 6, 6, 7, 7, 7, 7, 8, 8, 8, 9, 11}
$\bar{x} = 6.9$ from exercise 2-19'

(a)

x	x-$\bar{x}$	(x-$\bar{x}$)²
4	-2.9	8.41
5	-1.9	3.61
5	-1.9	3.61
6	-0.9	0.81
6	-0.9	0.81
6	-0.9	0.81
7	0.1	0.01
7	0.1	0.01
7	0.1	0.01
7	0.1	0.01
8	1.1	1.21
8	1.1	1.21
8	1.1	1.21
9	2.1	4.41
11	4.1	16.81
sum	+0.5*	42.95

$s^2 = 42.95/14 = 3.0679$

$s^2 = \underline{3.1}$

* The 0.5 is due to round-off error in $\bar{x} = 6.9$.

(b)

x	x^2
4	16
5	25
5	25
6	36
6	36
6	36
7	49
7	49
7	49
7	49
8	64
8	64
8	64
9	81
11	121
104	764

$SS(x) = \Sigma x^2 - \{(\Sigma x)^2/n\}$

$SS(x) = 764 - (104)(104)/15$

$s^2 = 42.9333/14 = 3.0667 = \underline{3.1}$

(c) $s = \sqrt{3.0667} = 1.751 = \underline{1.8}$

2-33

Machine	n	Σx	Σx^2
1	5	10.000	20.000030
2	5	10.000	20.000218

$\bar{x}_1 = 2.000$ and $s_1 = 0.0027$

$\bar{x}_2 = 2.000$ and $s_2 = 0.0074$

Even though both machines produced shafts of the same mean diameter, machine 2 had a standard deviation which was over 2.5 times the standard deviation of machine 1. Machine 2 will produce more shafts which do not meet specifications.

2-35

x	f	xf	x^2f
0	1	0	0
1	3	3	3
2	8	16	32
3	5	15	45
4	3	12	48
sum	20	46	128

$SS(x) = 128 - (46)(46)/20 = 22.2$

$s^2 = 22.2/19 = 1.168 = \underline{1.2}$

$s = \sqrt{1.168} = 1.081 = \underline{1.1}$

2-39

x	f	xf	$x^2 f$
17	4	68	1156
22	14	308	6776
27	19	513	13851
32	7	224	7168
37	5	185	6845
42	3	126	5292
47	2	94	4418
52	1	52	2704
sum	55	1570	48210

SS(x) = 48210 - (1570)(1570)/55 = 3393.63636

s^2 = 3393.63636/54 = 62.845

s = $\sqrt{62.845}$ = 7.927 = <u>7.9</u>

2-41
(a) The original numbers:
n = 6 Σx = 37,116 Σx^2 = 229,710,344

SS(x) = 229,710,344 - (37,116)(37,116)/6
      = 110,768.0

s^2 = 110,768.0/5 = <u>22,153.6</u>

(b) The smaller numbers:
n = 6 Σx = 1,116 Σx^2 = 318,344

SS(x) = 318,344 - (1,116)(1,116)/6
      = 110,768.0

s^2 = 110,768.0/5 = <u>22,153.6</u>

2-43   The statement must be incorrect. The standard deviation can never be negative. The computations need to be checked, they have to be in error.

2-45   Ranked data:
2.6, 2.7, 3.4, 3.6, 3.7, 3.9, 4.0, 4.4, 4.8, 4.8,
4.8, 5.0, 5.1, 5.6, 5.6, 5.6, 5.8, 6.8, 7.0, 7.0

(a) nk/100 = (20)(25)/100 = 5.0, i = 5.5th

Q_1 = (3.7 + 3.9)/2 = <u>3.8</u>

nk/100 = (20)(75)/100 = 15.0, i = 15.5th

Q_3 = (5.6 + 5.6)/2 = <u>5.6</u>

(b) midquartile = $(Q_1 + Q_3)/2$ = (3.8 + 5.6)/2

midquartile = <u>4.7</u>

(c) 1) $nk/100 = (20)(15)/100 = 3.0$, $i = 3.5$th

$P_{15} = (3.4 + 3.6)/2 = \underline{3.5}$

2) $nk/100 = (20)(33)/100 = 6.6$, $i = 7$th

$P_{33} = \underline{4.0}$

3) $nk/100 = (20)(90)/100 = 18.0$, $i = 18.5$th

$P_{90} = (6.8 + 7.0)/2 = \underline{6.9}$

2-46

(a) $Q_1 = P_{25}$, therefore $k = 25$ and $nk/100 = (40)(25)/100 = 10$. Thus, $i = 10.5$th and $Q_1 = \underline{8.3}$

(b) $Q_2 = P_{50}$, therefore $k = 50$ and $nk/100 = (40)(50)/100 = 20$. Thus, $i = 20.5$th and $Q_2 = (9.1+9.4)/2 = \underline{9.25}$

(c) $Q_3 = P_{75}$, therefore $k = 75$ and $nk/100 = (40)(75)/100 = 30$. Thus, $i = 30.5$th and $Q_3 = (10.7+11.0)/2 = \underline{10.85}$

(d) P_{95}, therefore $k = 95$ and $nk/100 = (40)(95)/100 = 38$. Thus, $i = 38.5$th and $P_{95} = (14.7+14.9)/2 = \underline{14.8}$

(e)
```
 7.1      8.3      9.25      10.85      15.5
  L       Q₁        X̃         Q₃         H
```

(f)
```
        ┌────┬──────┐
────────┤    │      ├────────────────────
        └────┴──────┘
 7.1   8.3  9.25  10.85                 15.5
```

2-49

$z = (x - \text{mean})/\text{st.dev.}$

(a) for $x = 54$, $z = (54-50)/4.0 = \underline{1.0}$

(b) for $x = 50$, $z = (50-50)/4.0 = \underline{0.0}$

(c) for $x = 59$, $z = (59-50)/4.0 = \underline{2.25}$

(d) for $x = 45$, $z = (45-50)/4.0 = \underline{-1.25}$

2-50

If $z = (x-\text{mean})/\text{st.dev.}$, then $x = (z)(\text{st.dev}) + \text{mean}$

For $z = 1.8$, $x = (1.8)(100) + 500 = \underline{680}$

2-53

For A: $z = (85-72)/8 = 1.625$
For B: $z = (93-87)/5 = 1.2$

Therefore, <u>A has the higher relative position</u>

2-55   (a) Since almost all the data fall between $\bar{x} - 3s$ and $\bar{x} + 3s$, then $(\bar{x} + 3s) - (\bar{x} - 3s) = 6s$ should be approximately equal to the range.

(b) Since 6s and the range are approximately equal, then s could be approximated by <u>range divided by 6.</u>

2-56   (a) at most 11%   (b) at most 6.25%

2-57

x	f	xf	x²f
1	7	7	7
2	10	20	40
3	22	66	198
4	8	32	128
5	7	35	175
6	2	12	72
7	3	21	147
8	0	0	0
9	1	9	81
sum	60	202	848

(b) $\bar{x} = 202/60 = 3.367 = \underline{3.4}$

$SS(x) = 848 - 202^2/60 = 167.9333$
$s = \sqrt{167.9333/59} = \sqrt{2.8463} = 1.687 = \underline{1.7}$

(c) $\bar{x} - s = \underline{1.7}$ and $\bar{x} + s = \underline{5.1}$
(d)   <u>47</u>, <u>78%</u>

(e) $\bar{x} - 2s = \underline{0.0}$ and $\bar{x} + 2s = \underline{6.8}$
(f)   <u>56</u>, <u>93%</u>

(g) $\bar{x} - 3s = \underline{-1.7}$ and $\bar{x} + 3s = \underline{8.5}$
(h)   <u>98.3%</u>

(i)   93% is at least 75% and 98.3% is at least 89%, which agrees with the claims of Chebyshev's Theorem.

(j)   78%, 93% and 98.3% are not approximately equal to the 68%, 95% and 99.7% claimed by the Empirical Rule.

2-60
(a)

Class limits	f	lower boundary	cum rel freq
1 - 3	6	0.5	0.12
4 - 6	9	3.5	0.30
7 - 9	8	6.5	0.46
10 - 12	10	9.5	0.66
13 - 15	6	12.5	0.78
16 - 18	4	15.5	0.86
19 - 21	4	18.5	0.94
22 - 24	2	21.5	0.98
25 - 27	1	24.5	1.00
	50		

(b)

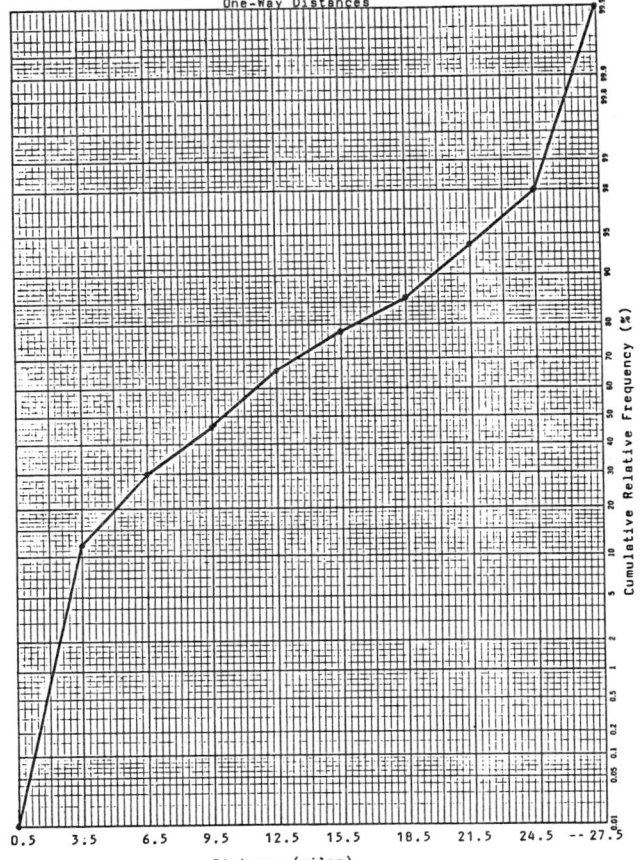

(c) Yes. The line segments (disregarding the end segments) form a path that is nearly a straight. This straight line appearance is what indicates the near normality of the distribution.

(d) $P_1 = \underline{3.41}$, $Q_1 = \underline{5.7}$, $\bar{X} = \underline{10.1}$,
 $Q_3 = \underline{14.7}$, $P_9 = \underline{19.7}$

2-61 (a) The mean increases, the sum increases when one data increases.

(b) The median is unchanged, the value of the median is affected only by the middle value(s).

(c) The mode is unchanged.

(d) The midrange increases, an increase in the value of one of the extremes increases the sum, H +L.

(e) The range increases, the difference between the lowest valued data and the highest has increased.

-11-

(f, g)　The variance and the standard deviation have both increased since the data are now more spread out.

2-63　Ranked data:
190, 215, 222, 226, 234, 242, 245, 249, 254, 258, 259, 265

(a)　$i = (12 + 1)/2 = 6.5$th
median $= (242 + 245)/2 = \underline{243.5}$

(b)　$nk/100 = (12)(75)/100 = 9.0$　$i = 9.5$th
$Q_3 = (254 + 258)/2 = \underline{256}$

(c)　$\bar{x} = 2859/12 = \underline{238.25}$

2-65　(a)　median = <u>11.8%</u>

(b)　mean $= 83.3/7 = \underline{11.9\%}$

(c)
x	x-x̄	|x - x̄|
8.6	-3.3	3.3
9.8	-2.1	2.1
10.9	-1.0	1.0
11.8	-0.1	0.1
13.7	1.8	1.8
13.9	2.0	2.0
14.6	2.7	2.7
		13.0

mean absolute deviation = $13.0/7 = 1.857 = \underline{1.9}$

2-68　Hours of TV watched

```
0. 00000 00000 0
0. 55
1. 0000
1. 55
2. 0000
2. 55555
3. 0
3. 
4. 0
4. 
5. 0
5. 
6. 0
```

(b)　mean $= 46.5/32 = 1.453 = \underline{1.45}$

(c)　median $= \underline{1}$,　$i = (32+1)/2 = 16.5$th

(d)　mode $= \underline{0}$

(e)　midrange $= (0+6)/2 = \underline{3}$

(f)　The mode represents the most common amount of television watched.

(g) The mean includes the concept of total amount of television watched by the people in the sample.

(h) range = 6 - 0 = <u>6</u>

(i) $SS(x) = 142.25 - (46.5^2/32) = 74.6797$
$s^2 = 74.68/31 = 2.409 = \underline{2.41}$

(j) $s = \sqrt{2.409} = 1.552 = \underline{1.55}$

2-70 Data: 63 67 66 63 69   74 72 70 71 71
             72 70 75 85 84   85 85 86 94 91
             90 90 95 105 104
$\Sigma x = 1,997$ $\Sigma x^2 = 163,205$

$\bar{x} = 1997/25 = \underline{79.88}$

$SS(x) = 163,205 - (1997^2/25) = 3684.64$

$s = \sqrt{3684.64/24} = \sqrt{153.5267} = 12.3906 = \underline{12.4}$

2-72

x	f
3	1
4	2
5	3
6	4
7	5
8	5
9	6
10	7
11	8
12	9
13	6
14	2
sum	58

(b) <u>58 litters</u>

(c) f is the number of litters, 58

2-74
(a)

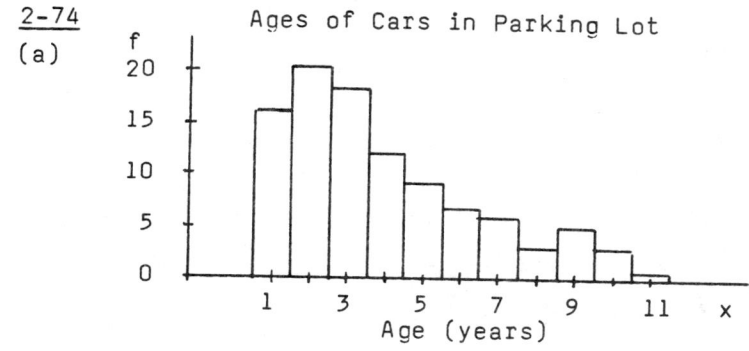
Ages of Cars in Parking Lot

-13-

(b)   mean = 397/100 = 3.97 = <u>4.0</u>

median = <u>3</u>, i = (100+1)/2 = 50.5th

mode = <u>2</u>

midrange = (1+11)/2 = <u>6</u>

midquartile = (2 + 5.5)/2 = <u>3.75</u>

(c) Q_1 <u>2</u>, nk/100 = (100)(25)/100 = 25, i = 25.5th

Q_3 <u>5.5</u>, nk/100 = (100)(75)/100 = 75, i = 75.5th

(d) P_{15} = <u>1</u>   nk/100 = (100)(15)/100 = 15, i = 15.5th

P_{12} = <u>1</u>   nk/100 = (100)(12)/100 = 12, i = 12.5th

(e)

x	f	xf	x^2f
1	16	16	16
2	20	40	80
3	18	54	162
4	12	48	192
5	9	45	225
6	7	42	252
7	6	42	294
8	3	24	192
9	5	45	405
10	3	30	300
11	1	11	121
sum	100	397	2239

range = 11 - 1 = <u>10</u>

SS(x) = 2239 - (397^2/100) = 662.91

s^2 = 662.91/99 = 6.696 = <u>6.7</u>

s = $\sqrt{6.696}$ = 2.588 = <u>2.6</u>

2-77   mean = $\Sigma x/n$, therefore 40 = $\Sigma x/75$ or Σx = (40)(75)
Σx = <u>3000</u>

s^2 = SS(x)/n-1, therefore SS(x) = (n-1)s^2 or

SS(x) = 100(74) = 7400

SS(x) = $\Sigma x^2 - (\Sigma x)^2/n$

Thus 7400 = $\Sigma x^2 - (3000^2/75)$ or

Σx^2 = 7,400 + 120,000 = <u>127,400</u>

2-79

Class limits	f	rel freq	cum freq
0 - 8	17	0.142	17
9 - 17	14	0.117	31
18 - 26	10	0.083	41
27 - 35	14	0.117	55
36 - 44	10	0.083	65
45 - 53	16	0.133	81
54 - 62	9	0.075	90
63 - 71	11	0.092	101
72 - 80	8	0.067	109
81 - 89	11	0.092	120
	120		

(a)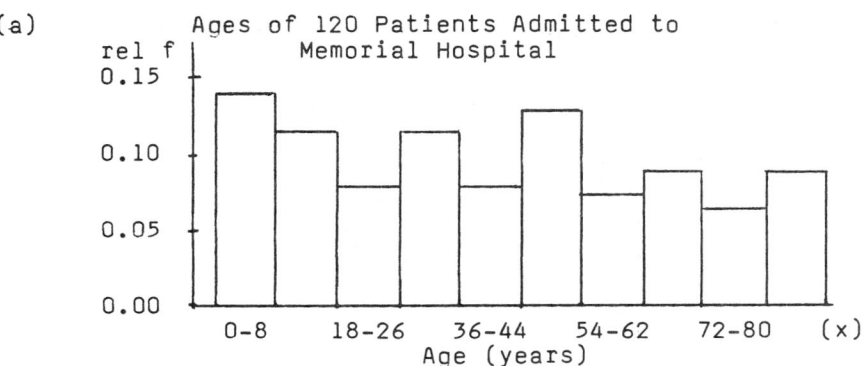

(b)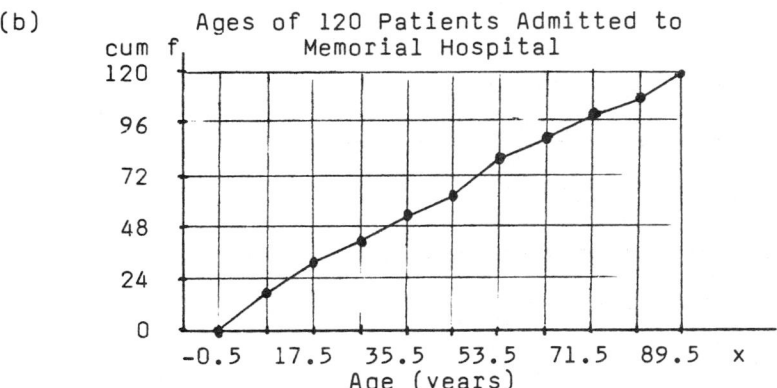

(c,d)

Class limits	f	x	xf	$x^2 f$
0 - 8	17	4	68	272
9 - 17	14	13	182	2366
18 - 26	10	22	220	4840
27 - 35	14	31	434	13454
36 - 44	10	40	400	16000
45 - 53	16	49	784	38416
54 - 62	9	58	522	30276
63 - 71	11	67	737	49379
72 - 80	8	76	608	46208
81 - 89	11	85	935	79475
	120		4890	280686

$\bar{x} = 4890/120 = \underline{40.75}$

$SS(x) = 280686 - (4890^2/120) = 81418.5$
$s = \sqrt{81418.5/119} = \sqrt{684.189} = 26.157 = \underline{26.2}$

2-81   Summary of data: n = 100, Σx = 1315

(a) $\bar{x}$ = 1315/100 = <u>13.15</u>

(b)   median = <u>13.85</u>,   i = (100+1)/2 = 50.5th

(c)   mode = <u>15.0</u>

(d)   midrange = (10.1 + 15.8)/2 = <u>12.95</u>

(e)   range = 15.8 - 10.1 = <u>5.7</u>

(f) Q_1 = <u>10.95</u>, nk/100 = (100)(25)/100 = 25,

       i = 25.5th

       Q_3 = <u>14.9</u>, nk/100 = (100)(75)/100 = 75,

       i = 75.5th

(g)   midquartile = (10.95 + 14.9)/2 = <u>12.925</u>

(h) P_{35} = <u>12.05</u>, nk/100 = (100)(35)/100 = 35,

       i = 35.5

       P_{64} = <u>14.5</u>, nk/100 = (100)(64)/100 = 64, i = 64.5

(i)
Class limits	x	f	xf	$x^2 f$
10.0 - 10.5	10.25	15	153.75	1575.9375
10.6 - 11.1	10.85	11	119.35	1294.9475
11.2 - 11.7	11.45	7	80.15	917.7175
11.8 - 12.3	12.05	3	36.15	435.6075
12.4 - 12.9	12.65	6	75.90	960.1350
13.0 - 13.5	13.25	2	26.50	351.1250
13.6 - 14.1	13.85	14	193.90	2685.5150
14.2 - 14.7	14.45	10	144.50	2088.0250
14.8 - 15.3	15.05	27	406.35	6115.5675
15.4 - 15.9	15.65	5	78.25	1224.6125
	sum	100	1314.80	17649.1900

(j)
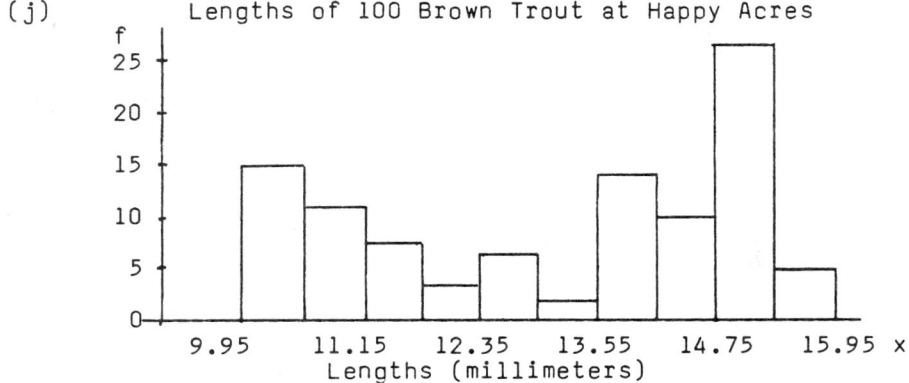

(k)

Class limits	x	f	cum rel f
10.0 - 10.5	10.25	15	0.15
10.6 - 11.1	10.85	11	0.26
11.2 - 11.7	11.45	7	0.33
11.8 - 12.3	12.05	3	0.36
12.4 - 12.9	12.65	6	0.42
13.0 - 13.5	13.25	2	0.44
13.6 - 14.1	13.85	14	0.58
14.2 - 14.7	14.45	10	0.68
14.8 - 15.3	15.05	27	0.95
15.4 - 15.9	15.65	5	1.00
	sum	100	

(l)

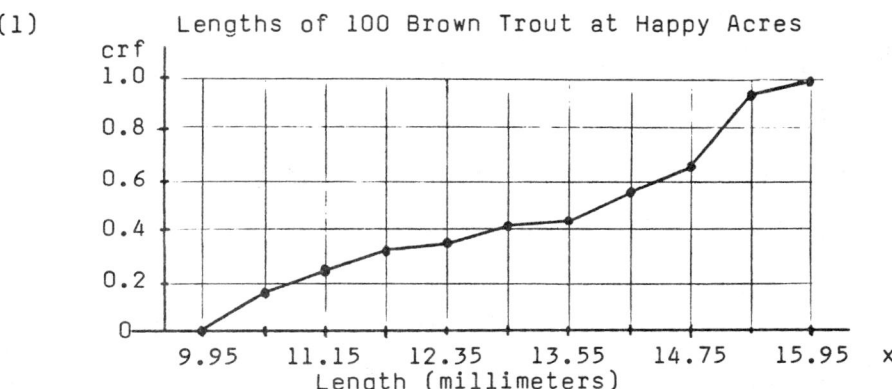

Lengths of 100 Brown Trout at Happy Acres

(m) $\bar{x}$ = 1314.80/100 = 13.148 = __13.15__

(n) SS(x) = 17649.1900 - (1314.80²/100) = 362.1996

   s = $\sqrt{362.1996/99}$ = $\sqrt{3.65858}$ = 1.9127 = __1.91__

2-85 98th, since there is approximately 97.5% of the distribution to the left of the z-score +2.

2-91 The empirical rule states that 99.7% of a normal distribution lies within three stadard deviations of the mean, that is between z = -3 and z = 3.

2-92 The interval 22,500 to 37,500 represents the mean plus or minus three standard deviations.

(a) If the distribution is normal, then __approximately 99.7%__ of the distribution is contained within the interval.

(b) If nothing is known about the shape of the distribution, we can be sure that __at least 89%__ of the distribution is contained within the interval.

Chapter Three

3-2

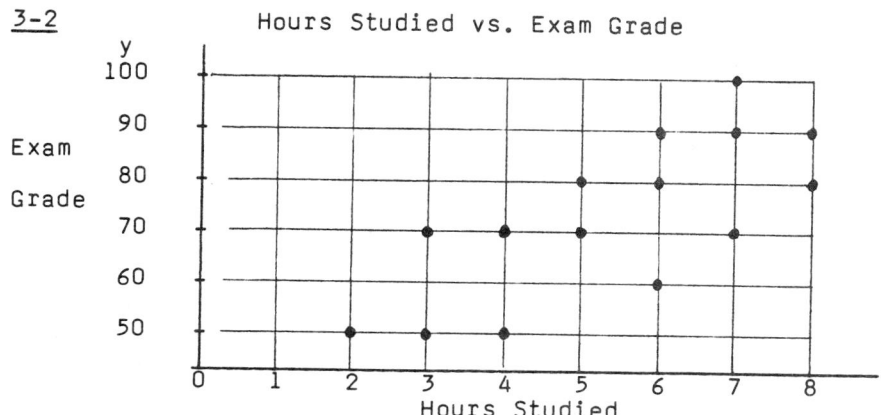

3-5

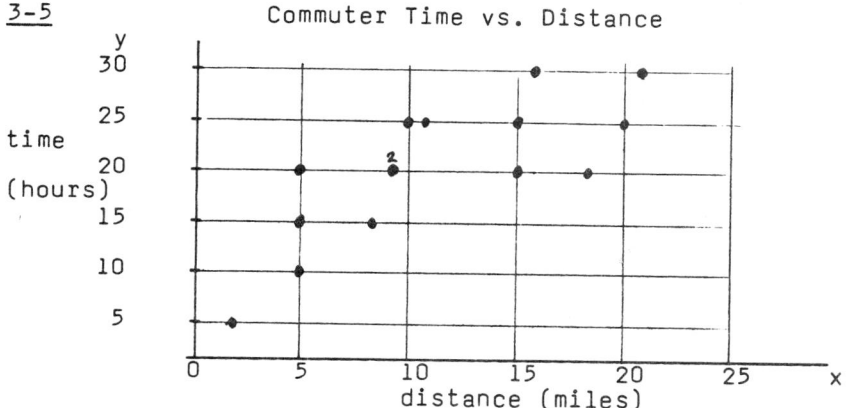

3-7 Impossible. The correlation coefficient must be a value between -1.0 and +1.0.

3-9 (a) Estimate r = 0.75

(b)

Data	x	y	x^2	xy	y^2
1	2	5	4	10	25
2	3	5	9	15	25
3	3	7	9	21	49
4	4	5	16	20	25
5	4	7	16	28	49
6	5	7	25	35	49
7	5	8	25	40	64
8	6	6	36	36	36
9	6	9	36	54	81
10	6	8	36	48	64
11	7	7	49	49	49
12	7	9	49	63	81
13	7	10	49	70	100
14	8	8	64	64	64
15	8	9	64	72	81
sum	81	110	487	625	842

$SS(x) = 487 - (81^2/15) = 49.6$

$SS(y) = 842 - (110^2/15) = 35.333$

$SS(xy) = 625 - \{(81)(110)/15\} = 31.0$

$r = 31.0/\sqrt{(49.6)(35.333)} = 0.74051 = \underline{0.741}$

3-12   Summations from extensions table:
$n = 16$, $\Sigma x = 2,564$, $\Sigma y = 1,238$, $\Sigma x^2 = 416,008$, $\Sigma xy = 199,438$, $\Sigma y^2 = 96,836$

$SS(x) = 416,008 - (2,564^2/16) = 5127$

$SS(y) = 96,836 - (1,238^2/16) = 1045.75$

$SS(xy) = 199,438 - \{(2,564)(1,238)/16\} = 1048.5$

$r = 1048.5/\sqrt{(5127)(1045.75)} = 0.45282 = \underline{0.453}$

3-14   (a) A moderate correlation might be interpreted to mean there appears to be a slight linear pattern showing that communities with higher viewership rates tend to also have higher rates of reported rape.

(b)   No cause-and-effect is shown by this information. It only says that when one rate is higher, the other tends to be higher, also.

3-16   Numerator of formula (3-1):
numerator $= \Sigma(x-\bar{x})(y-\bar{y})$

$= \Sigma(xy - \bar{x}y - x\bar{y} + \bar{x}\bar{y})$

$= \Sigma xy - \bar{x}\Sigma y - \bar{y}\Sigma x + n\bar{x}\bar{y}$

$= \Sigma xy - (\frac{\Sigma x}{n})\Sigma y - (\frac{\Sigma y}{n})\Sigma x + n(\frac{\Sigma x}{n})(\frac{\Sigma y}{n})$

$= \Sigma xy - \frac{\Sigma x \Sigma y}{n}$

$= SS(xy)$

Denominator of formula (3-1):
denominator $= (n-1)s_x s_y$

$= (n-1)\sqrt{\frac{SS(x)}{n-1}}\sqrt{\frac{SS(y)}{n-1}}$

$= \sqrt{SS(x)\ SS(y)}$

Therefore, formula (3-1) is equivalent to formula (3-2)

3-17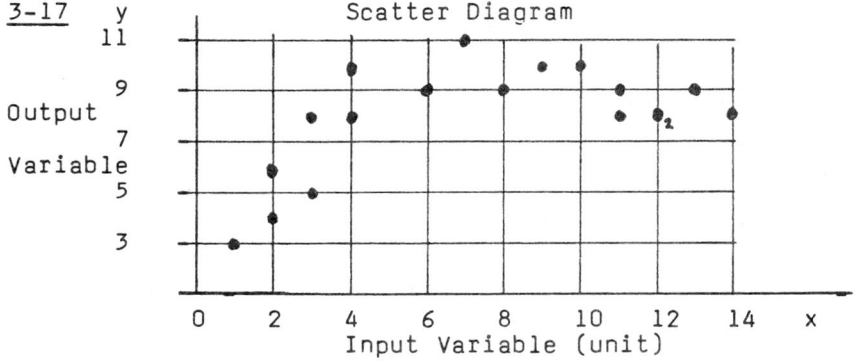

The above scatter diagram does not suggest a linear
relationship between the two variables. Therefore,
we would not be justified in using techniques of linear
regression. If we were to calculate the line of best
fit, we would get an equation. It would be worthless,
however.

3-18   The vertical scale shown on figure 3-21 is located at
x = 50 and is therefore not the y-axis. The y-axis is
the vertical line located at x = 0.

3-19 $\hat{y} = 7.31 - 0.01(50) = 6.81$

predicted value $= 6.81(10,000) = \underline{68,100}$

3-21(a)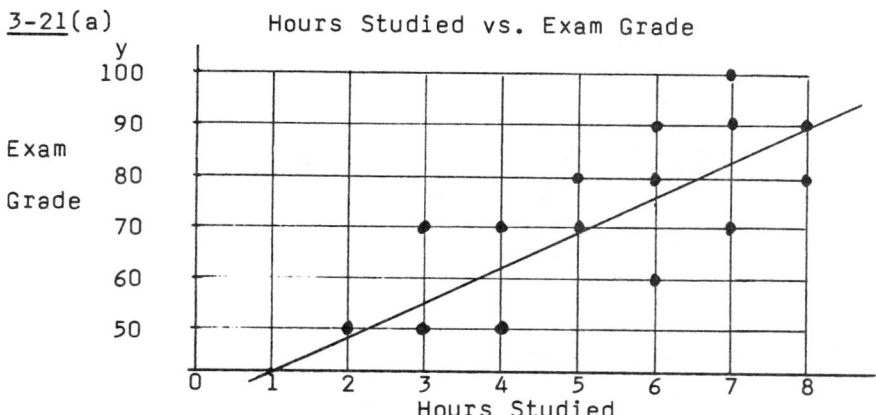

Estimation: $\hat{y} = 33 + 7x$

(b) From exercise 3-9: n = 15, Σx = 81, Σy = 110,
SS(x) = 49.6, SS(xy) = 31.0

b_1 = 31.0/49.6 = 0.625

$b = \frac{1}{15}\{110 - (0.625)(81)\}$ = 3.9583 = 3.96

$\hat{y} = 3.96 + 0.625x$, which translates to $\hat{y} = 39.6 + 6.25x$

(d) If x = 6, then one would expect y to be approximately equal to 39.6 + 6.25(6) = 77.1 or 77

(e) $\hat{y} = 77$ is the average score expected for all those who studied for 6 hours.

3-23 Summary of data: n = 10, Σx = 491, Σy = 19
Σx^2 = 30,221, Σxy = 1,196, Σy^2 = 49

(a) SS(x) = 30,221 - (491^2/10) = 6112.9

SS(y) = 49 - (19^2/10) = 12.9

SS(xy) = 1,196 - {(491)(19)/10} = 263.1

r = 263.1/$\sqrt{(6112.9)(12.9)}$ = 0.93692 = 0.937

(b) b_1 = 263.1/6112.9 = 0.043

$b_0 = \frac{1}{10}\{19 - (0.043)(491)\}$ = -0.21

$\hat{y} = -0.21 + 0.043x$

3-27 (a) The purpose of a correlation analysis is to determine whether two variables are linearly related or not. The product of correlation is the numerical value of r.

(b) The purpose of regression analysis is to determine the equation of the line of best fit. The product of regression is the equation.

3-31 (a) $x = 6{,}000$, $\hat{y} = 350 + 0.3(6000) = \underline{2150.00}$

$x = 12{,}000$, $\hat{y} = 350 + 0.3(12000) = \underline{3950.00}$

(b) These extreme values of x are outside the domain of the study. Therefore the answers obtained, 350 and 30,500, respectively, are unrealistic.

3-34 (a)

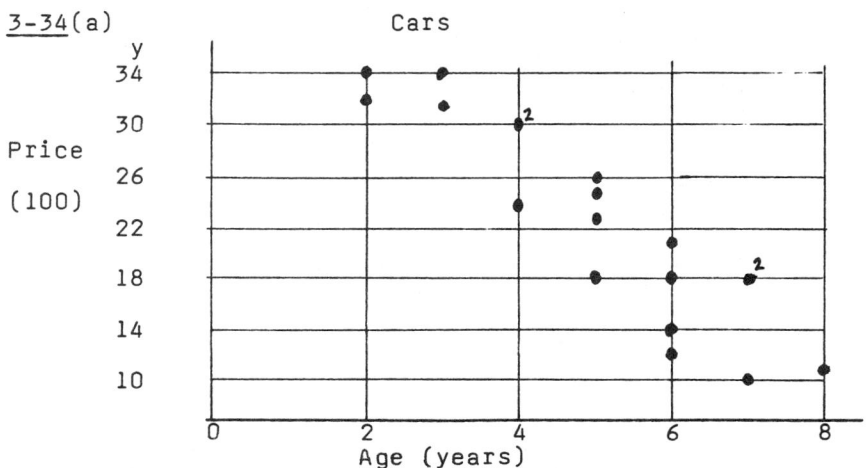

(b) Summations from extensions table:
$n = 19$, $\Sigma x = 95$, $\Sigma y = 429.5$, $\Sigma x^2 = 529$, $\Sigma xy = 1924.5$, $\Sigma y^2 = 10832.25$

$SS(x) = 529 - (95^2/19) = 54.0$

$SS(y) = 10832.25 - (429.5^2/19) = 1123.2895$

$SS(xy) = 1924.5 - (95)(429.5)/19 = -223.0$

$r = -223.0/\sqrt{(54.0)(1123.2895)} = 0.9054 = \underline{0.905}$

(c) <u>Yes</u>, there appears to be linear correlation between the variable.

(d) $b_1 = -223.0/54.0 = -4.1296 = -4.13$

$b_0 = \frac{1}{19}\{429.5 - (-4.1296)(95)\} = 43.253$

$\underline{\hat{y} = 43.25 - 4.13x}$

(e) On scatter diagram part (a).

(f) $\hat{y} = 43.25 - 4.13(5) = \underline{22.60\text{ hundred dollars}}$

Chapter Four

4-4 (a) Relative frequency for 1 - 0.14, 2 - 0.22, 3 - 0.12, 4 - 0.18, 5 - 0.22, 6 - 0.12

(b) Relative frequency for H - 0.50 and for T - 0.50

4-10 The 31.4 is a percentage of TV households tuned to Family Ties the week of the poll. This is an observed relative frequency, the proportion of the sampled households.

4-11 Let J = jack, Q = queen, K = king, S = spade, D = diamond, H = heart, C = club

S = {JS, JD, JH, JC, QS, QD, QH, QC, KS, KD, KH, KC}

4-14 (a) S = {HH, HT, TH, TT}

(b) S = {HH, HT, TH, TT}

Note: The sample spaces listing the possible outcomes are identical. The sample spaces listed above do not indicate relative probabilities.

4-16

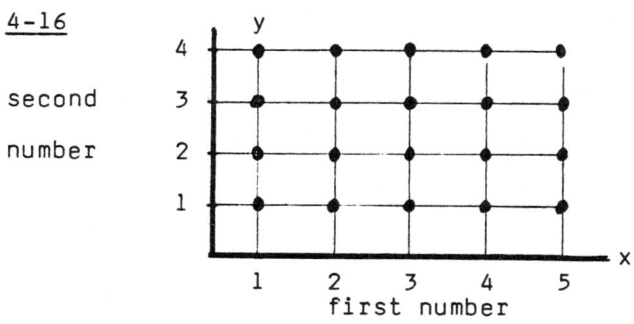

second number

4-18 D = defective, N = non-defective

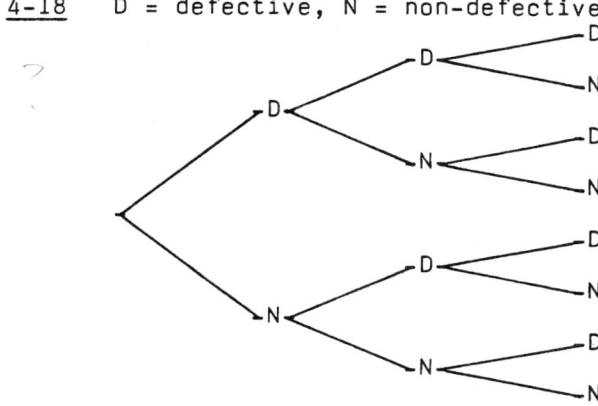

4-19 $P(R) = x$, $P(Y) = x$ and $P(G) = 2x$

$x + x + 2x = 4x = 1$ and therefore $x = 1/4$

$P(R) = 0.25$, $P(Y) = 0.25$, $P(G) = 0.50$

4-23   The three success ratings (highly successful, successful, and not successful) appear to be non-intersecting and their union appears to be the sample space. If this is true, none of the three sets of probabilities are appropriate.

(a)   A has a total probability of 1.2. The total must be exactly 1.0.

(b)   B has a negative probability of -0.1. All probability numbers are values between 0 and 1.

(c)   C has a total probability of 0.9. The total must be exactly 1.0.

4-25   (a) 0.55   (b) 0.40

4-27   (a) Not mutually exclusive. One head belongs to both events.

(b)   Not mutually exclusive. All sales that exceed 1000 also exceed 100, thus 1200 belongs to both events.

(c)   Not mutually exclusive. The student selected could be both male and over 21.

(d)   Mutually exclusive. The total cannot be both less than 7 and more than 7 at the same time, therefore there is no intersection.

4-29   (a) $P(\bar{A}) = 1 - 0.3 = \underline{0.7}$

(b) $P(\bar{B}) = 1 - 0.4 = \underline{0.6}$

(c) $P(A \text{ or } B) = 0.3 + 0.4 = \underline{0.7}$

(d) $P(A \text{ and } B) = \underline{0.0}$ (Mutually exclusive means there is no intersection.)

4-31   No. Female students can be working students. Also, if the given probabilities are correct, there must be an intersection otherwise the total probability is more than 1.0.

4-33   (a) $75/200 = \underline{0.375}$

(b) $(115/200) + (130/200) - (75/200) = 170/200 = \underline{0.85}$

(c) $P(\bar{B}) = 1 - P(B) = 1 - (130/200) = 70/200 = \underline{0.35}$

4-36 (a) independent   (b) not independent
      (c) independent   (d) independent
      (e) not independent (If you do not own a car, your car can not have a flat tire.)
      (f) not independent

4-39 (a) Using formula (4-7b):

      $0.20 = 0.4 \times P(A|B)$, therefore $P(A|B) = \underline{0.5}$

(b)   Using formula (4-7a):

      $0.20 = 0.3 \times P(B|A)$, Therefore $P(B|A) = \underline{0.666...}$

(c)   No, A and B are not independent events.

4-41 (a) $P(A) = 12/52 = 3/13$ and $P(A|B) = 6/26 = 3/13$
      Therefore, A and B are <u>independent</u> events.

(b) $P(A) = 12/52 = 3/13$ and $P(A|C) = 3/13$
      Therefore, A and C are <u>independent</u> events.

(c) $P(B) = 26/52 = 1/2$ and $P(B|C) = 1$
      Therefore, A and B are <u>dependent</u> events.

NOTE: B independent of A and C independent of A <u>does not</u> imply that B and C are independent.

4-43 $P(\text{all three Red}) = P(R_1) P(R_2|R_1) P(R_3|R_1 \text{ and } R_2)$

(a)   with replacement:
      $P(\text{all red}) = (4/7)(4/7)(4/7) = \underline{64/343}$

(b)   without replacement:
      $P(\text{all red}) = (4/7)(3/6)(2/5) = \underline{4/35}$

4-46 Let D = a defective is drawn and N = non-defective is drawn on either the first(1) or second(2) drawing.

(a)   P(both are defective) =
      $= P(D_1 \text{ and } D_2)$
      $= P(D_1)P(D_2|D_1)$
      $= (3/25)(2/24) = \underline{0.01}$

(b)   P(exactly one defective) =
      $= P((D_1 \text{ and } N_2) \text{ or } (N_1 \text{ and } D_2))$
      $= P(D_1 \text{ and } N_2) + P(N_1 \text{ and } D_2)$
      $= (3/25)(22/24) + (22/25)(3/24) = \underline{0.22}$

(c)   P(neither is defective) =
      $= P(N_1 \text{ and } N_2)$

      $= P(N_1)P(N_2|N_1)$

      $= (22/25)(21/24) = \underline{0.77}$

4-49 (a) $P(G|H) = 0.1/0.4 = \underline{0.25}$

(b) $P(H|G) = 0.1/0.5 \ \underline{0.2}$

(c) $P(\overline{H}) = 1 - 0.4 = \underline{0.6}$

(d) $P(G \text{ or } H) = 0.5 + 0.4 - 0.1 = \underline{0.8}$

(e) $P(G \text{ or } \overline{H}) = 0.5 + 0.6 - 0.4 = \underline{0.7}$

(f) No. They intersect, therefore P(intersection) can not be equal to 0.0.

(g) No. $P(G|H) \ne P(G)$

4-51 (a) $P(M \text{ and } N) = \underline{0.0}$ (mutually exclusive)

(b) $P(M \text{ or } N) = 0.3 + 0.4 = \underline{0.7}$

(c) $P(M \text{ or } \overline{N}) = P(\overline{N}) = 1 - 0.4 = \underline{0.6}$

(d) $P(M|N) = \underline{0.0}$ (mutually exclusive)

(e) $P(M|\overline{N}) = 0.3/0.6 = \underline{0.5}$

(f) No. Mutually exclusive events are disjoint, therefore dependent.

4-53 (a) P(A wins on 1st turn) = $1/2$

P(B wins on 1st turn) = $1/4$

P(C wins on 1st turn) = $1/8$

(b) P(A wins) = $1/2 + 1/16 = \underline{9/16}$

P(B wins) = $1/4 + 1/32 = \underline{9/32}$

P(C wins) = $1/8 + 1/64 = \underline{9/64}$

4-56 (a) $P(A \text{ and } B) = \underline{0.2}$ (b) $P(A \text{ or } C) = \underline{0.7}$
(c) $P(A|C) = 0.2/0.4 = \underline{0.5}$

4-59 Let F_i represent factory where shoe was produced, with $i = 1, 2, 3$. M represents mislabeled.

$M = (M \text{ and } F_1) \text{ or } (M \text{ and } F_2) \text{ or } (M \text{ and } F_3)$
$P(M) = P(F_1)P(M|F_1) + P(F_2)P(M|F_2) + P(F_3)P(M|F_3)$
$= (0.25)(0.01) + (0.60)(0.005) + (0.15)(0.02)$
$= \underline{0.0085}$

4-61 (a) and (b) are true. These two statements are examples of the law of large numbers.

4-63 (a) 89/216 (b) 111/216 (c) 64/113

4-65 (a) False. For mutually exclusive events the or event is found by adding the two probabilities.

(b) True. 0.2 + 0.5 - (0.2)(0.5) = 0.6

(c) False. For mutually exclusive events, there is no intersection, therefore P(R and S) would be 0.0.

(d) False. 0.2 + 0.5 = 0.7, not 0.6

4-68 Let A_i represent a 6 on the ith roll

P(6 occurs first on the 5th roll) =

= P(A_1 and A_2 and A_3 and A_4 and A_5)
= P(A_1)P(A_2)P(A_3)P(A_4)P(A_5)
= (5/6)(5/6)(5/6)(5/6)(1/6)
= 0.080

4-70 (a) P(A and B) = P(B)×P(A|B) = (0.36)(0.88) = 0.3168

(b) P(B|A) = P(A and B)/P(A) = 0.3168/0.68 = 0.4659

(c) No, P(A) ≠ P(A|B)

(d) No, P(A and B) ≠ 0

(e) It would mean that the two events candidate wants job and RJB wants candidate could both not happen.

4-74 (a) 6/12 = 0.5 (b) 1/6 = 0.167 (c) 5/12 = 0.417

4-77 (a) P(seedless) = 0.30

(b) P(white) = 0.60

(c) P(pink and seedless) = 0.10

(d) P(pink or seedless) = 0.10 + 0.20 + 0.30 = 0.60

(e) P(pink|seedless) = 0.10/0.30 = 0.333

(f) P(seedless|pink) = 0.10/0.40 = 0.25

4-80 Let A = thunderstorm in vicinity of airport and B = plane lands on time.

Given P(A) = 0.70 and P(B|A) = 0.80

P(A and B) = P(A)P(B|A) = (0.70)(0.80) = 0.56

4-84   P(satisfactory) = P(all good) = p^6

(a)   P(satisfactory|p=0.9) = 0.9^6 = 0.531

(b)   P(satisfactory|p=0.8) = 0.8^6 = 0.262

(c)   P(satisfactory|p=0.6) = 0.6^6 = 0.047

4-87

```
                       0.8   alarm    0.9   wakes on time*
              sets  <       rings <
              alarm                   0.1   does not wake
         0.7 /      0.2               0.2   wakes on time*
           /              does not <
          <               ring        0.8   does not wake
           \
         0.3 \                  0.2   wakes on time*
              forgets     <
              alarm        0.8   does not wake
```

P(wakes on time) = (0.7)(0.8)(0.9) + (0.7)(0.2)(0.2) + (0.3)(0.2) = 0.504 + 0.028 + 0.060 = 0.592

4-88   Identify the proofreaders as A and B

(a)   P(error is found) =

P{(A finds error on page read) or (B finds error)} =

P{(error on pg and A finds it) or (error on pg and B finds it)} =

(0.5)(0.8) + (0.5)(0.8) = 0.80

(b)   P(error is found|both read both pages) =

P(A finds error or B finds error) =

0.8 + 0.8 - (0.8)(0.8) = 0.96

(c)   P(error is found|both read one randomly selected pg)=

P{(error is on pg A reads and A finds it) or

    (error is on pg B reads and B finds it)} =

{(0.5)(0.8)} + {(0.5)(0.8)} - {(0.5)(0.8)}{(0.5)(0.8)}

= 0.64

4-90   Let: A = person selected is from age group 0 - 20
            B = person selected is from age group 21 - 40
            C = person selected is from age group above 40
            G = person had an abnormal glucose tolerance test

P(G) = P(A and G) + P(B and G) + P(C and G)
     = P(A)P(G|A) + P(B)P(G|B) + P(C)P(G|C)
     = (0.3)(0.05) + (0.4)(0.04) + (0.3)(0.10)
     = 0.061

Chapter Five

5-1 The random variable is the <u>number of children</u> per family.

The possible values for the random variable are 0, 1, 2, 3, ... , n, where n is the maximum number of children for any family in the community.

5-3 x = n(bulls-eye shots)
x = 0, 1, 2, 3, ..., 7, 8

5-5

x	0	1	2	3
P(x)	0.20	0.30	0.40	0.10

Notice that each P(x) is a value between 0 and 1, and that the sum of all the P(x) values is exactly 1.0.

5-8

x	P(x)
1	4/10
2	3/10
3	2/10
4	1/10
sum	10/10 = 1.0

It is a probability function:

(i) each P(x) is a value between zero and one,

(ii) the sum of the P(x)s is one.

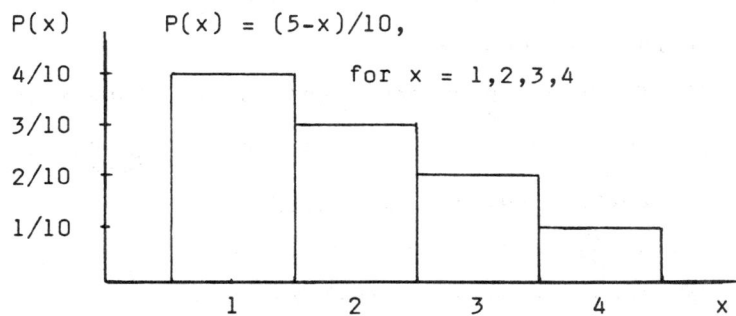

P(x) = (5-x)/10, for x = 1,2,3,4

5-10

x	R(x)
0	0.2
1	0.2
2	0.2
3	0.2
4	0.2
sum	1.0

It is a probability function:

(a) each P(x) is a value between zero and one,

(b) the sum of the P(x)s is one.

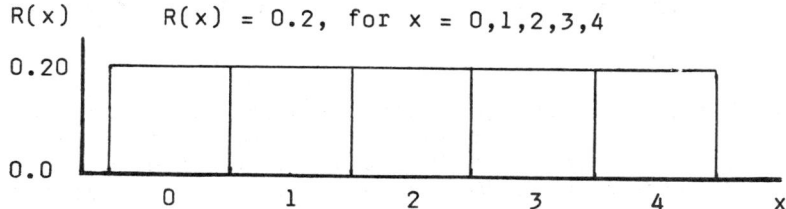

R(x) = 0.2, for x = 0,1,2,3,4

5-12 (a) The percentages are based on observed data. The accidents reported for 7,458 athletes, not all athletes.

(b) Using <u>age</u> as the variable and <u>skiing</u> as the identifying characteristic for the population of athletes, the percentage of injured for each age becomes the probability. Each injured person can only belong to one age class.

(c) One injured person could have multiple responses to the <u>anatomic site of injury</u> variable. For example, one person could injure a knee, an ankle and break a leg all in one accident.

5-13

x	P(x)	xP(x)	$x^2 P(x)$
1	4/10	4/10	4/10
2	3/10	6/10	12/10
3	2/10	6/10	18/10
4	1/10	4/10	16/10
	10/10 = 1.0	20/10 = 2.0	50/10 = 5.0

$\mu = \underline{2.0}$

$\sigma^2 = 5.0 - (2.0)^2$
$\sigma = \underline{1.0}$

5-15

x	P(x)	xP(x)	$x^2 P(x)$
0	0.1	0.0	0.0
1	0.2	0.2	0.2
2	0.4	0.8	1.6
3	0.2	0.6	1.8
4	0.1	0.4	1.6
	1.0	2.0	5.2

$\mu = \underline{2.0}$

$\sigma^2 = 5.2 - (2.0)^2 = 1.2$
$\sigma = \underline{1.1}$

5-18 P(x) = 0.1, x = 0,1,2,3, ... ,9

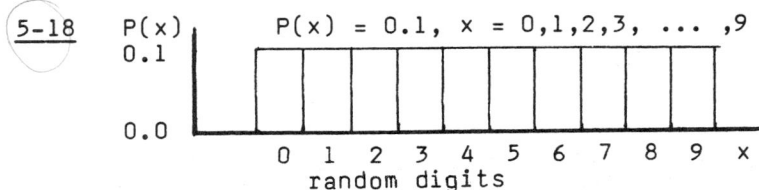

continued

x	P(x)	xP(x)	$x^2 P(x)$
0	0.1	0.0	0.0
1	0.1	0.1	0.1
2	0.1	0.2	0.4
3	0.1	0.3	0.9
4	0.1	0.4	1.6
5	0.1	0.5	2.5
6	0.1	0.6	3.6
7	0.1	0.7	4.9
8	0.1	0.8	6.4
9	0.1	0.9	8.1
	1.0	4.5	28.5

$\mu = \underline{4.5}$ $\sigma^2 = 28.5 - (4.5)^2 = 8.25$
$\sigma = \underline{2.87}$

5-19 $\sigma^2 = \Sigma(x - \mu)^2 P(x)$

$\Sigma(x^2 - 2x + \mu^2)P(x)$

$\Sigma x^2 P(x) - 2\mu\Sigma xP(x) + \mu^2 \Sigma P(x)$

$\Sigma x^2 P(x) - 2\mu^2 + \mu^2$

$\sigma^2 = \Sigma x^2 P(x) - \mu^2$ or $\Sigma x^2 P(x) - \{\Sigma xP(x)\}^2$

5-20 (a) $4! = 4\times3\times2\times1 = \underline{24}$

(b) $7! = 7\times6\times5\times4\times3\times2\times1 = \underline{5,040}$ (c) $0! = \underline{1}$

(d) $\frac{6\times5\times4\times3\times2\times1}{2\times1} = \underline{360}$ (e) $\frac{5\times4\times3\times2\times1}{3\times2\times1\times2\times1} = \underline{10}$

(f) $\frac{6\times5\times4\times3\times2\times1}{4\times3\times2\times1\times2\times1} = \underline{15}$ (g) $0.3\times0.3\times0.3\times0.3 = \underline{0.0081}$

(h) $\frac{7\times6\times5\times4\times3\times2\times1}{3\times2\times1\times4\times3\times2\times1} = \underline{35}$ (i) $\frac{5\times4\times3\times2\times1}{2\times1\times3\times2\times1} = \underline{10}$

(j) $\frac{3\times2\times1}{1\times3\times2\times1} = \underline{1}$ (k) $4\times0.2\times0.8\times0.8\times0.8 = \underline{0.4096}$

(l) $1\times1\times(0.7)^5 = \underline{0.16807}$

5-21 <u>x is not a binomial random variable</u> because the trials are not independent. The probability of success(get an ace) changes from trial to trial. On the first trial it is 4/52. The probability of an ace on the second trial can be either 4/51 or 3/51 depending on whether the first trial was a success or not. The probability of success on the third and fourth trials depends on what has occurred on the previous trials. The probability of success does not remain constant throughout the experiment since we do not have independence of trials.

(b) <u>x is a binomial random variable</u> because the trials are independent. The probability of success on any one trial will be 1/13 since each card that is drawn is replaced before the next drawing. This way the probabilities remain constant trial after trial.

5-23 (a)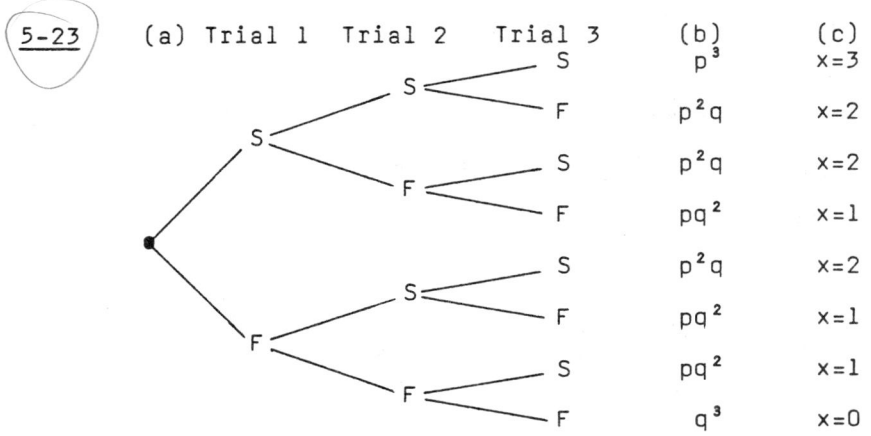

Trial 1	Trial 2	Trial 3	(b)	(c)
		S	p^3	x=3
	S	F	p^2q	x=2
S		S	p^2q	x=2
	F	F	pq^2	x=1
		S	p^2q	x=2
F	S	F	pq^2	x=1
		S	pq^2	x=1
	F	F	q^3	x=0

(e) $P(x) = \binom{3}{x} p^x q^{3-x}$, x = 0, 1, 2, 3

5-25 (a) $\binom{4}{1}(0.3)^1(0.7)^3 = \underline{0.4116}$
(b) $\binom{3}{2}(0.8)^2(0.2)^1 = \underline{0.384}$
(c) $\binom{2}{0}(0.25)^0(0.75)^2 = \underline{0.5625}$
(d) $\binom{5}{2}(1/3)^2(2/3)^3 = \underline{0.329218}$
(e) $\binom{4}{2}(0.5)^2(0.5)^2 = \underline{0.375}$
(f) $\binom{3}{3}(1/6)^3(5/6)^0 = \underline{0.0046296}$

5-26 (a) 0.001   (b) 0.279   (c) 0.031
(d) 0.886   (e) 0.002   (f) 0.057
(g) 0+ represents a probability of less than 0.0005.

5-28 (a) Yes. n = 5, p = P(H) = 1/2, q = P(T) = 1/2, each toss is a trial.

(b) $P(x) = \binom{5}{x}(1/2)^x(1/2)^{5-x}$, for x = 0,1,...,5

(c) See exercise 5-27.

5-30 The number of defective items should be fairly small and therefore easier to count.

5-33 P(shut down) = P(x > 2), where x represents the number of defective in the sample of size 10.

$$P(x > 2) = 1 - \{P(x = 0) + P(x = 1)\}$$

$$P(x=0) = \binom{10}{0}(0.005)^0(0.995)^{10} = 0.9511$$

$$P(x=1) = \binom{10}{1}(0.005)^1(0.995)^9 = 0.0478$$

$$P(x > 2) = 1 - (0.9511 + 0.0478)$$
$$= \underline{0.0011}$$

5-37 $(q + p)^3 = q^3 + 3q^2p + 3qp^2 + p^3$

$P(0) = q^3 \qquad P(1) = 3q^2p$

$P(2) = 3qp^2 \qquad P(3) = p^3$

5-38

x	P(x)	xP(x)	$x^2P(x)$
0	1/32	0/32	0/32
1	5/32	5/32	5/32
2	10/32	20/32	40/32
3	10/32	30/32	90/32
4	5/32	20/32	80/32
5	1/32	5/32	25/32
sum	32/32	80/32	240/32

$\mu = 80/32 = \underline{2.5}$

$\sigma = \sqrt{240/32 - (80/32)^2} = \sqrt{1.25} = 1.118 = \underline{1.2}$

(b) $\mu = (5)(1/2) = \underline{2.5}$

$\sigma = \sqrt{(5)(1/2)(1/2)} = \sqrt{1.25} = 1.118 = \underline{1.2}$

(c) The answers are the same by both methods.

5-39 (a) $\mu = np = (100)(1/13) = 7.692 = \underline{7.7}$

$\sigma = \sqrt{npq} = \sqrt{(100)(1/13)(12/13)} = 2.665 = \underline{2.7}$

(b) $\mu = np = (400)(0.06) = \underline{24.0}$

$\sigma = \sqrt{npq} = \sqrt{(400)(0.06)(0.94)} = 4.7497 = \underline{4.7}$

(c) $\mu = np = (50)(0.88) = \underline{44.0}$

$\sigma = \sqrt{npq} = \sqrt{(50)(0.88)(0.12)} = 2.298 = \underline{2.3}$

(d) x = n(defective), p = P(defective) = 0.02
$\mu = np = (125)(0.02) = \underline{2.5}$

$\sigma = \sqrt{npq} = \sqrt{(125)(0.02)(0.98)} = 1.565 = \underline{1.6}$

5-40 We know that $\mu = 200 = np$ and $\sigma = 10 = \sqrt{npq}$

By squaring the standard deviation formula and by combining this information, we have

$200q = 100$, from which we find $q = 0.5$.

Since $p = 1-q$, $p = \underline{0.5}$ (the other one-half).

Returning to original information, $np = 200$,

$n(0.5) = 200$, from which we find $n = \underline{400}$.

5-43 (i) Each probability, P(x), has a value between zero and one.

(ii) The sum of the individual probabilities is exactly one.

5-45 (a)

x	9	10	11	12
f(x)	0.25	0.25	0.25	0.25

<u>Yes</u> it is a probability function. Each f(x) is a value between 0 and 1, and the f(x) values sum to exactly 1.0.

(b)

x	1	2	3	4
f(x)	1.0	0.5	0.0	-0.5

<u>No</u> it is not a probability function. Even though the f(x) values sum to exactly 1.0, f(4) is not a value between 0 and 1.

(c)

x	0	1	2	3
f(x)	1/25	3/25	7/25	13/25

<u>No</u> it is a probability function. Each f(x) is a value between 0 and 1, however the f(x) values sum to 24/25, which is not 1.0.

5-47

x	P(x)	xP(x)
0	0.40	0.00
1	0.30	0.30
2	0.25	0.50
3	0.05	0.15
sum	1.00	0.95

$\mu = xP(x) = \underline{0.95}$

(b) P(at least 1 mistake) = P(x = 1, 2, 3)
    = P(1) + P(2) + P(3) = $\underline{0.6}$

5-51 $\sigma^2 = \Sigma x^2 P(x) - \mu^2$
$100 = \Sigma x^2 P(x) - 2500$ or $\Sigma x^2 P(x) = \underline{2600}$

5-53 $\binom{4}{0} = \frac{4 \times 3 \times 2 \times 1}{1 \times 4 \times 3 \times 2 \times 1} = 1$, 1 branch - no success

$\binom{4}{1} = \frac{4 \times 3 \times 2 \times 1}{1 \times 3 \times 2 \times 1} = 4$, 4 branches - one success

$\binom{4}{2} = \frac{4 \times 3 \times 2 \times 1}{2 \times 1 \times 2 \times 1} = 6$, 6 branches - two successes

$\binom{4}{3} = \frac{4 \times 3 \times 2 \times 1}{3 \times 2 \times 1 \times 1} = 4$, 4 branches - three successes

$\binom{4}{4} = \frac{4 \times 3 \times 2 \times 1}{4 \times 3 \times 2 \times 1 \times 1} = 1$, 1 branch - four successes

```
trial 1  trial 2  trial 3  trial 4    outcome   x-value
                              S         SSSS     x=4
                    S
                              F         SSSF     x=3
            S
                              S         SSFS     x=3
                    F
                              F         SSFF     x=2
   S
                              S         SFSS     x=3
                    S
                              F         SFSF     x=2
            F
                              S         SFFS     x=2
                    F
                              F         SFFF     x=1

                              S         FSSS     x=3
                    S
                              F         FSSF     x=2
            S
                              S         FSFS     x=2
                    F
                              F         FSFF     x=1
   F
                              S         FFSS     x=2
                    S
                              F         FFSF     x=1
            F
                              S         FFFS     x=1
                    F
                              F         FFFF     x=0
```

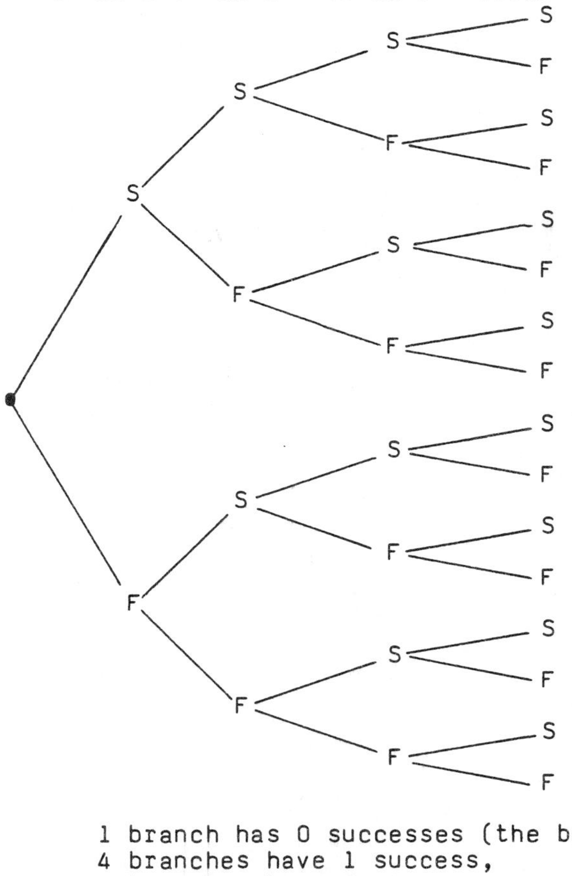

1 branch has 0 successes (the bottom branch),
4 branches have 1 success,
6 branches have 2 successes,
4 branches have 3 successes,
1 branch has 4 successes.

5-54 $n = 4$, $x = n(\text{males})$, $p = 2/3$, $q = 1/3$

$P(x) = \binom{4}{x}(2/3)^x (1/3)^{4-x}$, for $x = 0,1,2,3,4$

x	0	1	2	3	4
P(x)	0.012	0.099	0.296	0.395	0.198

5-57 $P(x=3,4 \text{ or } 5|n=5, p=0.25) = P(3) + P(4) + P(5)$

$P(x=3) = \binom{5}{3}(0.25)^3(0.75)^2 = 0.0879$

$P(x=4) = \binom{5}{4}(0.25)^4(0.75)^1 = 0.0146$

$P(x=5) = \binom{5}{5}(0.25)^5(0.75)^0 = 0.0010$

$P(x=3,4 \text{ or } 5) = 0.0879 + 0.0146 + 0.0010 = \underline{0.1035}$

5-60 $P(x=8,9 \text{ or } 10|n=10, p=0.90) = 0.194 + 0.387 + 0.349$
$= \underline{0.930}$

5-62 $P(x=8,9 \text{ or } 10|n=11, p=0.90) = 0.071 + 0.213 + 0.384$
$= \underline{0.668}$

5-64 (a) $P(x=0,1 \text{ or } 2|n=10, p=0.10) = 0.349 + 0.387 + 0.194$
$= \underline{0.930}$

(b) $P(x=2,3,4,\ldots \text{ or } 10|n=10, p=0.10) = 1 - P(x=0,1)$
$= 1 - (0.349 + 0.387) = \underline{0.264}$

5-68 Two engine plane: $n = 2$, $p = 0.95$
$P(\text{successful flight}) = P(x = 1,2) = 0.095 + 0.902$
$= \underline{0.997}$

Four engine plane: $n = 4$, $p = 0.95$
$P(\text{successful flight}) = P(x = 2,3,4)$
$= 0.014 + 0.171 + 0.815$
$= \underline{1.000}$

The four engine plane has the higher probability of a successful flight.

5-70 Let success be the event that an item is defective on a given draw. The probability of success on the first draw is 3/10 or 0.3. The probability of success on the second draw is dependent on the results of the first draw. If the first was defective then the probability is 2/9. If the first is not defective, then the probability is 3/9. Since the trials are not independent, x is not a binomial random variable.

5-71 (a) $P(\text{accepted}) = P(x=0 \text{ or } 1|n=10, p=0.05)$
$= 0.599 + 0.315 = \underline{0.914}$

(b) $P(\text{not accepted}) = P(x=2,3 \text{ or } 10|n=10, p=0.20)$
$= 1 - P(x=0 \text{ or } 1|n=10, p=0.20)$
$= 1 - (0.107 + 0.268) = \underline{0.625}$

5-74 (a) $\binom{8}{2} = \underline{28}$

(b)
```
12  23  34  45  56  67  78
13  24  35  46  57  68
14  25  36  47  58
15  26  37  48
16  27  38
17  28
18
```

(c) P(each possible set) = $\underline{1/28}$

(d)

x	P(x)
3	1/28
4	1/28
5	2/28
6	2/28
7	3/28
8	3/28
9	4/28
10	3/28
11	3/28
12	2/28
13	2/28
14	1/28
15	1/28
	28/28

(e)

xP(x)	$x^2P(x)$
3/28	9/28
4/28	16/28
10/28	50/28
12/28	72/28
21/28	147/28
24/28	192/28
36/28	324/28
30/28	300/28
33/28	363/28
24/28	288/28
26/28	338/28
14/28	196/28
15/28	225/28
252/28	2520/28

$\mu = 252/28 = \underline{9.0}$

$\sigma^2 = 2520/28 - (252/28)^2 = \underline{9.0}$

$\sigma = \sqrt{9.0} = \underline{3.0}$

5-76
$\mu = \Sigma xP(x)$
$= (1)(\frac{1}{n}) + (2)(\frac{1}{n}) + \ldots + (n)(\frac{1}{n})$
$= (\frac{1}{n})(1 + 2 + \ldots + n)$
$= (\frac{1}{n})(\frac{(n)(n+1)}{2})$
$= \frac{n+1}{2}$

$\frac{40320}{2 \cdot 720} = 1440$

-37-

Chapter Six

6-1   A bell shaped distribution with a mean of zero and a standard deviation of one.

6-2   (a) 0.4032   (b) 0.3997   (c) 0.4993
      (d) 0.4761

6-3   (a) 0.4821   (b) 0.4949   (c) 0.3849
      (d) 0.4418

6-4   (a) <u>0.4394</u>

(b)   0.5000 - 0.4394 = <u>0.0606</u>

(c)   0.5000 + 0.4394 = <u>0.9394</u>

(d)   2(0.4394) = <u>0.8788</u>

6-7   (a) <u>0.5000</u>

(b)   0.5000 - 0.3531 = <u>0.1469</u>

(c)   0.5000 + 0.4893 = <u>0.9893</u>

(d)   0.4452 + 0.5000 = <u>0.9452</u>

(e)   0.5000 - 0.4452 = <u>0.0548</u>

6-10 (a) 1.14   (b) 0.47   (c) 1.66
      (d) 0.86   (e) 1.74   (f) 2.23

6-12 (a) 1.65   (b) 1.96   (c) 2.33

6-15 (a) +0.84   (b) +1.04

6-18 (a)

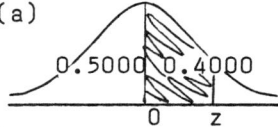

The z-score corresponding to an area of 0.4000 is 1.28

$z = 1.28$ corresponds P_{90}

(b)

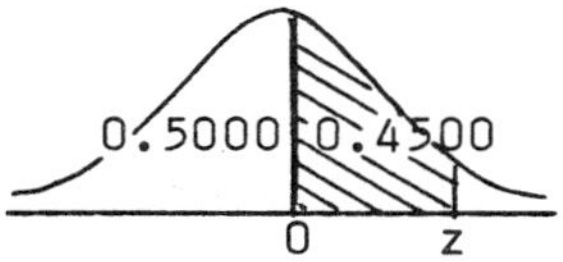

The z-score corresponding to an area of 0.4500 is 1.65

$z = 1.65$ corresponds P_{95}

(c)

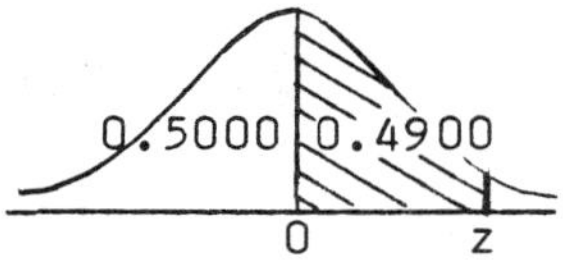

The z-score corresponding to an area of 0.4900 is 2.33

$z = 2.33$ corresponds P_{99}

6-19 (a) $P(x>60) = P(z>\frac{60-60}{10}) = P(z>0.00)$
$= 0.5000$

(b) $P(60<x<72) = P(0.00<z<\frac{72-60}{10})$
$= P(0.00<z<1.20) = 0.3849$

(c) $P(57<x<83) = P(\frac{57-60}{10}<z<\frac{83-60}{10})$
$= P(-0.30<z<2.30) = 0.1179 + 0.4893$
$= 0.6072$

(d) $P(65<x<82) = P(\frac{65-60}{10}<z<\frac{82-60}{10})$
$= P(0.50<z<2.20) = 0.4861 - 0.1915$
$= 0.2946$

(e) $P(38<x<78) = P(\frac{38-60}{10}<z<\frac{78-60}{10})$
$= P(-2.20<z<1.80) = 0.4861 + 0.4641$
$= 0.9502$

(f) $P(x<38) = P(z<\frac{38-60}{10}) = P(z<-2.20)$
$= 0.5000 - 0.4861 = 0.0139$

6-23 (a)

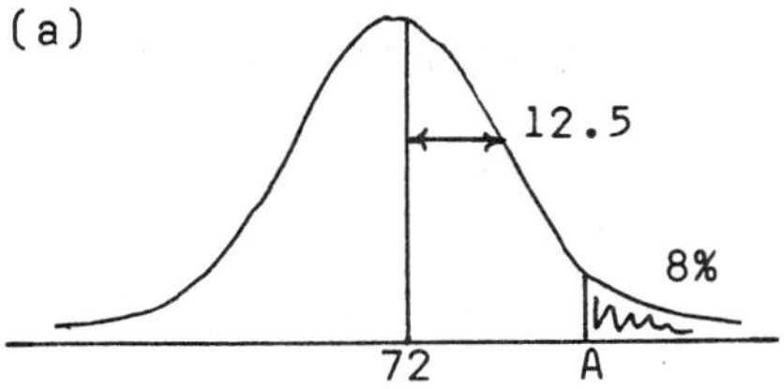

8% implies $z_A = 1.41$

$1.41 = \frac{A-72}{12.5}$

$A = (1.41)(12.5) + 72 = 89.625 = \underline{89.6}$

(b)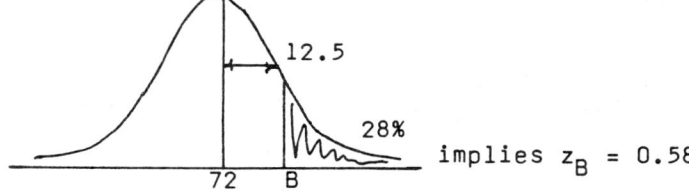

implies $z_B = 0.58$

$0.58 = \dfrac{B-72}{12.5}$

$B = (0.58)(12.5) + 72 = \underline{79.2}$

(c)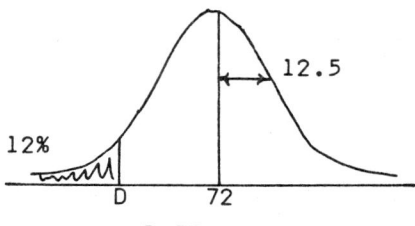

implies $z_D = -1.175$

$-1.175 = \dfrac{D-72}{12.5}$

$D = (-1.175)(12.5) + 72 = 57.3125 = \underline{57.3}$

6-26 (a) $P(x<2.0) = P(z<\dfrac{2.0-3.7}{1.4})$

$\qquad = P(z<-1.21)$
$\qquad = 0.5000 - 0.3869 = \underline{0.1131}$

(b) $P(x>6.0) = P(z>\dfrac{6.0-3.7}{1.4})$

$\qquad = P(z>1.64)$
$\qquad = 0.5000 - 0.4495 = \underline{0.0505}$

(c) P_{75} implies that $z = +0.67$

$0.67 = \dfrac{x - 3.7}{1.4}$

$x = (0.67)(1.4) + 3.7 = 4.638 = \underline{4.64 \text{ min.}}$

6-29 Let x = weight of a container.

We wish to determine σ so that $P(x < 15.8) = 0.05$

$z = -1.65$

$-1.65 = \dfrac{15.8-16.0}{\sigma}$

$\sigma = -0.2/-1.65 = \underline{0.12}$

6-32   (a) z(0.03)   (b) z(0.14)   (c) z(0.75)
        (d) z(0.13)   (e) z(0.91)   (f) z(0.82)

6-35   (a) 1.28, 1.65, 1.96, 2.05, 2.33, 2.58
        (b) -2.58, -2.33, -2.05, -1.96, -1.65, -1.28

6-36   (a) $z(0.95) = -1.65$ and $z(0.025) = 1.96$

The area between -1.65 and 1.96 is
$$0.95 - 0.025 = \underline{0.925}$$

(b) $z(0.025) - z(0.95) = 1.96 - (-1.65) = \underline{3.61}$

6-37   (a) $np = 3$ and $nq = 7$
The approximation is <u>not appropriate</u> since $np < 5$.

(b) $np = 0.5$ and $nq = 99.5$
The approximation is <u>not appropriate</u> since $np < 5$.

(c) $np = 50$ and $nq = 450$
The approximation is <u>appropriate</u> since both $np > 5$ and $nq > 5$.

(d) $np = 10$ and $nq = 40$
The approximation is <u>appropriate</u> since both $np > 5$ and $nq > 5$.

6-39 $P(x = 6) = P(5.5 < x < 6.5)$
$$= P\left(\frac{5.5-7.2}{\sqrt{2.88}} < z < \frac{6.5-7.2}{\sqrt{2.88}}\right)$$
$$= P(-1.00 < z < -0.41)$$
$$= 0.3413 - 0.1591 = \underline{0.1822}$$

$P(x = 6 | n=12, p=0.6) = \underline{0.177}$, from Table 4

6-41 $P(x \leq 8) = P(x < 8.5) = P\left(z < \frac{8.5-5.6}{\sqrt{3.36}}\right)$
$$= P(z < 1.58)$$
$$= 0.5000 + 0.4429 = \underline{0.9429}$$

$P(x \leq 8 | n=14, p=0.4) = \underline{0.943}$, from Table 4

6-43   Let x represent the number of patients in the 250 who will experience a side-effect. Then x is a binomial random variable with $\mu = np = 12.5$ and $\sigma = \sqrt{npq} = 3.45$

The normal approximation:

$P(x < 15.5) = P(z < (15.5-12.5)/3.45)$
$= P(z < 0.87) = 0.5000 + 0.3087 = \underline{0.8078}$

6-46 $P(x > 600 | n=1800, p=P(\text{drop out})=0.3) = P(x > 600.5)$
$\{\mu = 540, \sigma = 19.44\}$
$$= P\left(z > \frac{600.5-540.0}{19.44}\right) = P(z > 3.11)$$
$$= 0.5000 - 0.4991 = \underline{0.0009}$$

6-50

x	-7	-6	-5	-4	-3	-2	-1	0	1	2	3	4	5
f	4	2	2	4	2	9	18	24	39	55	48	52	62

6	7	8	9	10	11	12	13	14	15	16	17	18	19	20	21
46	40	34	22	22	9	13	9	1	2	6	2	2	0	2	1

(b) The interval $\bar{x} - s = 0.5$ to $\bar{x} + s = 9.1$ contains all x-values from 1 to 9.

398/532 = 0.748   or 74.8%

The interval $\bar{x} - 2s = -3.8$ to $\bar{x} + 2s = 13.4$ contains all x-values from -3 to 13.

482/532 = 0.906   or 90.6%

The interval $\bar{x} - 3s = -8.1$ to $\bar{x} + 3s = 17.7$ contains all x-values from -8 to 17.

505/532 = 0.949   or 94.9%

(c) Empirical rule: 68%,   95%,   99.7%
This data has: 74.8%, 90.6%, 94.9%   Not very close.

(d) The interval $\bar{x} = 4.8$ to $\bar{x} + 1.5s = 11.25$ contains all x-values from 5 to 11.

235/532 = 0.442   or 44.2%   compared to 43.32%

(e) The histogram looks normal, however not all of the percentages compare very closely.

6-52   (a) 1.26   (b) 2.16   (c) 1.13

6-55   (a) $P(|z| > 1.68) = P(z<-1.68) + P(z>1.68)$
             $= 2(0.5000 - 0.4535) = 2(0.0465) = \underline{0.0930}$

(b) $P(|z| < 2.15) = P(-2.15 < z < 2.15)$
             $= 2(0.4842) = \underline{0.9684}$

6-57   (a) 1.175 or 1.18   (b) 0.58
       (c) -1.04   (d) -2.33

6-59

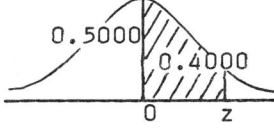

z = 1.28

$1.28 = \dfrac{P_{90} - 15.0}{3.0}$

$P_{90} = \underline{18.8}$

6-61

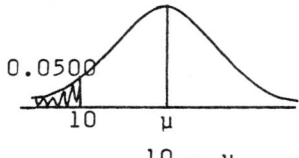

P_{90} implies $z = 1.28$

$1.28 = \dfrac{84 - 70}{\sigma}$

$\sigma = 14/1.28 = \underline{10.9}$

6-64

0.0500 below implies $z = -1.65$

$-1.65 = \dfrac{10 - \mu}{0.02}$

$\mu = 10.0 - (-1.65)(0.02) = \underline{10.033}$

6-67 (a) The normal approximation is reasonable since both $np = 7.5$ and $nq = 17.5$ are greater than 5.

(b) $\mu = np = (25)(0.3) = \underline{7.5}$

$\sigma = \sqrt{npq} = \sqrt{(25)(0.3)(0.7)} = \sqrt{5.25} = \underline{2.29}$

6-69 (a) $P(3 \text{ wrong in } 5) = P(x=3 | n=5, p=0.05) = \underline{0.001}$
(Table 4)

(b) $P(\text{no more than } 3) = P(x=0,1,2,3 | n=5, p=0.05)$
$= 0.774 + 0.204 + 0.021 + 0.001 = \underline{1.000}$
(Table 4)

(c) $P(\text{no more than } 3 \text{ in } 15) = P(x=0,1,2,3 | n=15, p=0.05)$
$= 0.463 + 0.366 + 0.135 + 0.031 = \underline{0.995}$
(Table 4)

(d) $P(\text{no more than } 3 \text{ in } 150) = P(x<3.5)$

$= P(z < \dfrac{3.5-7.5}{\sqrt{7.125}}) = P(z<-1.50)$

$= 0.5000 - 0.4332 = \underline{0.0668}$

6-71 (a) $P(x>45) = P(x>45.5) = P(z > \dfrac{45.5-42.5}{\sqrt{6.375}})$

$= P(z>1.19) = 0.5000 - 0.3830 = \underline{0.1170}$

(b) $P(40<x<50) = P(40.5<x<49.5)$

$= P(\dfrac{40.5-42.5}{\sqrt{6.375}} < z < \dfrac{49.5-42.5}{\sqrt{6.375}})$

$= P(-0.79<z<2.77) = 0.2852 + 0.4972 = \underline{0.7824}$

(c) $P(x<35) = P(x<34.5) = P(z < \dfrac{34.5-42.5}{\sqrt{6.375}})$

$= P(z<-3.17) = 0.5000 - 0.4992 = \underline{0.0008}$

6-73

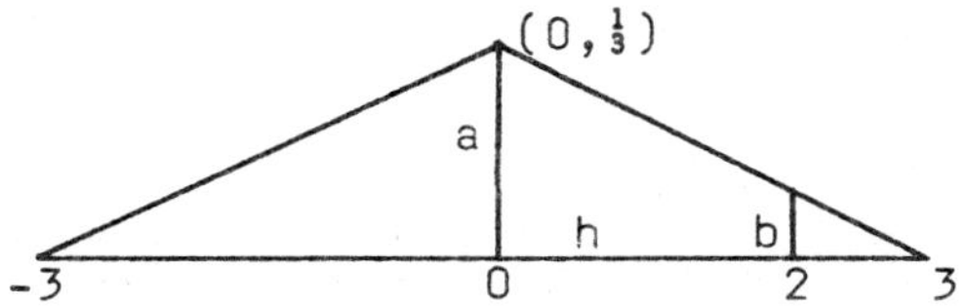

<pre>
              (0, 1/3)
              /|\
            a/ | \
            / h b\
         /_____|___|__\
        -3   0   2 3
</pre>

(a)   Area of triangle = $\frac{1}{2}$bh = $\frac{1}{2}$(6)($\frac{1}{3}$)

      Total area = <u>1.0</u>

(b)   The area between 0 and 2 is a trapezoid.
A = (1/2)(h)(a+b)

      h = 2,   a = 1/3,   b is the value of y at x=2

      b = (-1/9)(2) + (1/3) = 1/9

      Therefore, A = (1/2)(2)(1/3 + 1/9) = 4/9 = <u>0.4444</u>

(c)   <u>0.4772</u>

Chapter Seven

7-1 (a) A sampling distribution of sample means is the distribution formed by the means from all possible samples of a fixed size that can be taken from a population.

(b) It is one element in the distribution of the means of all samples of size 3.

(c) There were 25 equally likely samples. Therefore each has a probability of 1/25 or 0.04.

7-2 (a)
```
11  31  51  71  91
13  33  53  73  93
15  35  55  75  95
17  37  57  77  97
19  39  59  79  99
```

(b)
$\bar{x}$	1	2	3	4	5	6	7	8	9
$P(\bar{x})$	0.04	0.08	0.12	0.16	0.20	0.16	0.12	0.08	0.04

(c)
R	0	2	4	6	8
P(R)	0.20	0.32	0.24	0.16	0.08

7-5

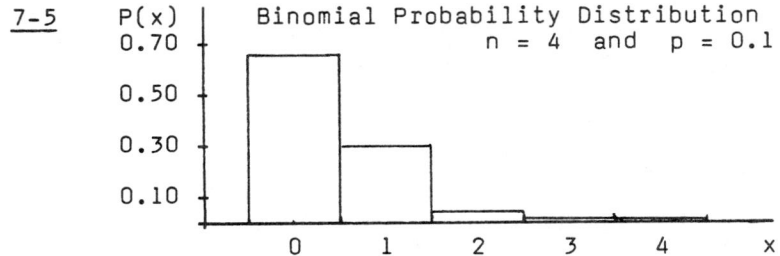

Binomial Probability Distribution n = 4 and p = 0.1

Notice that even though the population from which the samples are selected is highly skewed to the right, the histogram of sample means tends to be mound shape. That is, has some of the properties of a normal distribution.

7-7 (a) Each boss was asked to report the <u>average number of hours</u> each spends on phone per day.

(b)

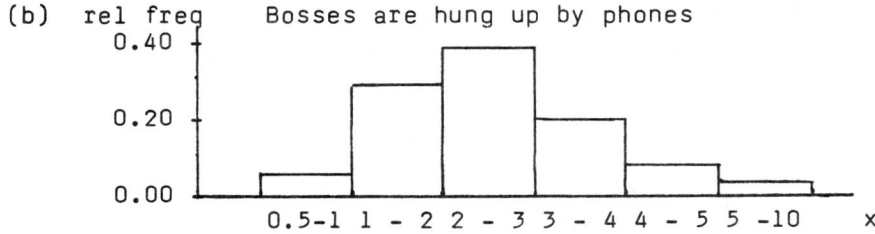

Bosses are hung up by phones

The shape is mounded. The shape shown in the histogram is some what deceiving since the classes are not of equal width.

(c) The <u>average</u> reported by each boss may not be a mean from a specific sample. Even if a boss did keep a record and calculate a sample mean for reporting, there was no control on the size of individual samples.

<u>7-8</u> (a) one
(b) $\sigma_{\bar{x}} = \sigma/\sqrt{n}$

<u>7-10</u> $25/\sqrt{16} = 6.25$ $25/\sqrt{25} = 5.00$
  $25/\sqrt{50} = 3.54$ $25/\sqrt{100} = 2.50$

<u>7-14</u> (a) $P(45<\bar{x}<55) = P(\frac{45-50}{10/\sqrt{36}} < z < \frac{55-50}{10/\sqrt{36}})$
  $= P(-3.00<z<3.00) = 2(0.4987) = \underline{0.9974}$

(b) $P(\bar{x}>48) = P(z > \frac{48-50}{10/\sqrt{36}})$
  $= P(z>-1.20) = 0.3849 + 0.5000 = \underline{0.8849}$

(c) $P(47<\bar{x}<53) = P(\frac{47-50}{10/\sqrt{36}} < z < \frac{53-50}{10/\sqrt{36}})$
  $= P(-1.80<z<1.80) = 2(0.4641) = \underline{0.9282}$

<u>7-15</u> x is a binomial random variable, n = 16 ans p = 0.5.
  $\mu = np = (16)(0.5) = 8.0$ and
  $\sigma = \sqrt{npq} = \sqrt{(16)(0.5)(0.5)} = \sqrt{4} = 2.0$
  $P(7.5<\bar{x}<8.5) = P(\frac{7.5-8.0}{2/\sqrt{16}} < z < \frac{8.5-8.0}{2/\sqrt{16}})$
  $= P(-1.25<z<1.25) = 2(0.3944) = \underline{0.7888}$

<u>7-17</u>
(a) $P(38<x<40) = P(\frac{38-39}{2} < z < \frac{40-39}{2})$
  $= P(-0.50<z<0.50) = 2(0.1915) = \underline{0.3830}$

(b) $P(38<\bar{x}<40) = P(\frac{38-39}{2/\sqrt{30}} < z < \frac{40-39}{2/\sqrt{30}})$
  $= P(-2.74<z<2.74) = 2(0.4969) = \underline{0.9938}$

(c) $P(x>40) = P(z > \frac{40-39}{2})$
  $= P(z>0.50) = 0.5000 - 0.1915 = \underline{0.3085}$

(d) $P(\bar{x}>40) = P(z > \frac{40-39}{2/\sqrt{30}})$
  $= P(z>2.74) = 0.5000 - 0.4969 = \underline{0.0031}$

7-20   Let x represent the total baggage weight for the 100 passengers. Then we wish to determine $P(\Sigma x > 2125)$, but that is the same as the $P(\bar{x} > 21.25)$

$$P(\bar{x} > 21.25) = P(z > \frac{21.25-20}{4/\sqrt{100}})$$
$$= P(z > 3.13)$$
$$= 0.5000 - 0.4991 = \underline{0.0009}$$

7-23   (a) Since n > 30, we know that $(\bar{x} - \mu)/\sigma_{\bar{x}}$ will be approximately normally distributed.

$$P(\mu - 2\sigma_{\bar{x}} < \bar{x} < \mu + 2\sigma_{\bar{x}})$$
$$= P(-2 < z < 2) = 2(0.4772) = \underline{0.9544}$$

(b) $P(\mu - 3\sigma_{\bar{x}} < \bar{x} < \mu + 3\sigma_{\bar{x}})$
$$= P(-3 < z < 3) = 2(0.4987) = \underline{0.9974}$$

7-24   (a) If n = 25, $\sigma_{\bar{x}} = \sigma/\sqrt{n} = 5/\sqrt{25} = 1.0$

$$P(\mu - 1\sigma_{\bar{x}} < \bar{x} < \mu + 1\sigma_{\bar{x}})$$
$$= P(-1 < z < 1) = 2(0.3413) = \underline{0.6826}$$

(b)   If n = 100, $\sigma_{\bar{x}} = \sigma/\sqrt{n} = 5/\sqrt{100} = 0.5$

$$P(\mu - 1\sigma_{\bar{x}} < \bar{x} < \mu + 1\sigma_{\bar{x}})$$
$$= P(-2 < z < 2) = 2(0.4772) = \underline{0.9544}$$

(c)   If n = 225, $\sigma_{\bar{x}} = \sigma/\sqrt{n} = 5/\sqrt{225} = 1/3$

$$P(\mu - 1\sigma_{\bar{x}} < \bar{x} < \mu + 1\sigma_{\bar{x}})$$
$$= P(-3 < z < 3) = 2(0.4987) = \underline{0.9974}$$

7-25   (a) If $\sigma = 5$, $\sigma_{\bar{x}} = \sigma/\sqrt{n} = 5/\sqrt{25} = 1.0$

$$P(\mu - 1\sigma_{\bar{x}} < \bar{x} < \mu + 1\sigma_{\bar{x}})$$
$$= P(-1 < z < 1) = 2(0.3413) = \underline{0.6826}$$

(b)   If $\sigma = 2.5$, $\sigma_{\bar{x}} = \sigma/\sqrt{n} = 2.5/\sqrt{25} = 0.5$

$$P(\mu - 1\sigma_{\bar{x}} < \bar{x} < \mu + 1\sigma_{\bar{x}})$$
$$= P(-2 < z < 2) = 2(0.4772) = \underline{0.9544}$$

(c)   If $\sigma = 10$, $\sigma_{\bar{x}} = \sigma/\sqrt{n} = 10/\sqrt{25} = 2$

$P(\mu - 1\sigma_{\bar{x}} < \bar{x} < \mu + 1\sigma_{\bar{x}})$

$= P(-0.5 < z < 0.5) = 2(0.1913) = \underline{0.3830}$

7-28 (a) Individual score x: the distribution of x's is normal with a mean of 720 and a standard deviation of 60.

(b)   Mean scores $\bar{x}$: The distribution of $\bar{x}$'s is normal with a mean of 720 and a standard deviation of 6 $(60/\sqrt{100})$.

(c) $P(x<725.6) = P(z<\frac{725.6-720.0}{60}) = P(z<0.09)$

$= 0.5000 + 0.0359 = \underline{0.5359}$

(d) $P(\bar{x}<725.6) = P(z<\frac{725.6-720.0}{60/\sqrt{100}}) = P(z<0.93)$

$= 0.5000 + 0.3238 = \underline{0.8238}$

7-30 $P(2.63-e<x<2.63+e) = 0.95$

$P(-1.96<z<+1.96) = 0.95$, using Table 5

$+1.96 = \frac{(2.63+e) - 2.63}{0.25}$

$e = \underline{0.49}$

7-33 $P(|\bar{x}-22.80|>0.75) = P(|z|>\frac{0.75}{3.00/\sqrt{200}})$

$= 2P(z>3.53) = 2(0.5000 - 0.4998) = \underline{0.0004}$

7-35 (a) $P(70<\bar{x}<75)$

$= P(\frac{70-72}{10/\sqrt{100}} < z < \frac{75-72}{10/\sqrt{100}})$

$= P(-2.00<z<3.00)$
$= 0.4772 + 0.4987 = \underline{0.9759}$

(b)   one standard error = $10/\sqrt{100} = 1.0$, therefore

  $\underline{69 \text{ to } 75}$

7-38 Using the manufacturers claims:
$\mu = 35,000$ and $\sigma = 5,000$, and sample size n = 100

$P(\bar{x} < 31,000) = P(z < \frac{31,000 - 35,000}{5,000/\sqrt{100}})$

$= P(z < -8.00) \quad \underline{0.0000+}$

Since the chances are virtually zero, the manufacturers figures of 35,000 and 5,000 should be in doubt.

Chapter Eight

8-1 (a) Type A correct decision: The accused is indeed innocent and is acquitted.

Type I error: The accused is infact innocent but is convicted.

Type II error: The accused is actually guilty but is acquitted.

Type B correct decision: The accused is guilty and is convicted.

(b) No (c) No

8-4 A type I error occurs if the company concludes that the additive increases average coverage when in fact it does not.

A type II error occurs if the company concludes that the additive does not increase the coverage when in fact it does increase the coverage.

8-6 We are willing to allow the type I error to occur with a probability of (a) 0.001, (b) 0.05, (c) 0.10

8-9 The null hypothesis states that the teaching techniques have no effect (change is equal to zero), while the alternative (what they want to show) is the statement that the teaching techniques do have an effect of student's exam scores. They are looking for improvement, therefore using a one-tail test.

8-13 (a) The critical region is the set of all values of the test statistic that will cause us to reject H_o.

(b) The critical value(s) is the value that forms the boundary between the critical region and the non-critical region. The critical value is in the critical region.

8-14 Because α and β are inter-connected. If we reduce α then β will become larger.

8-16 (a) $H_o: \mu = 26$ yrs($<,=$), $H_a: \mu > 26$
(b) $H_o: \mu = 36.7$ lbs($>,=$), $H_a: \mu < 36.7$
(c) $H_o: \mu = 1600$ hrs($>,=$), $H_a: \mu < 1600$
(d) $H_o: \mu = 210$ lbs($<,=$), $H_a: \mu > 210$
(e) $H_o: \mu = 4.7$ mi($>,=$), $H_a: \mu < 4.7$
(f) $H_o: \mu = 80$ deg, $H_a: \mu \neq 80$
(g) $H_o: \mu = 20$ deg($>,=$), $H_a: \mu < 20$
(h) $H_o: \mu = 570$ lbs/unit, $H_a: \mu \neq 570$

8-18 (a) 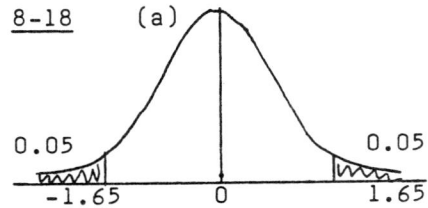 (b)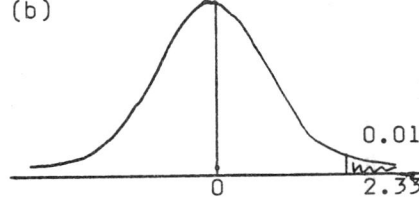

(c) 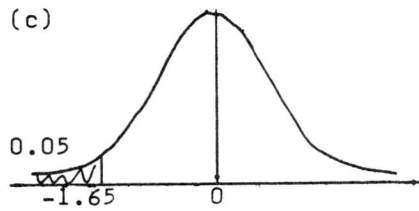 (d)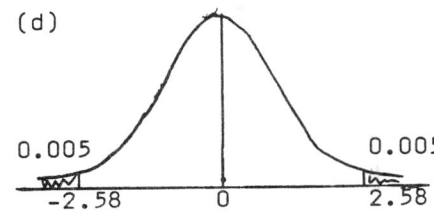

(e) 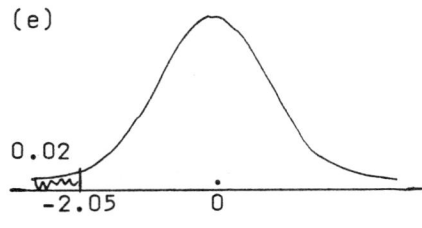 (f)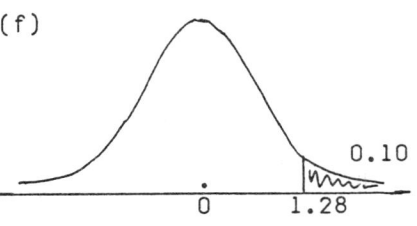

(g) 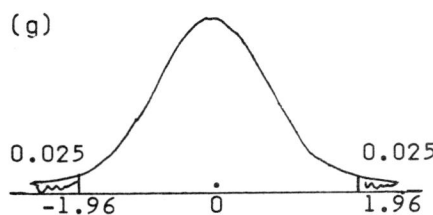 (h)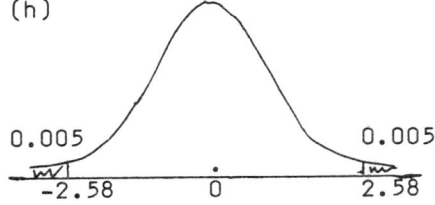

8-19 (a) For a sample of size n = 100, the standard error is $1.0/\sqrt{100} = 0.1$.

$\bar{x} = 4.8$ is $(4.8 - 4.5)/0.1 = 3$ standard errors above the mean $\mu = 4.5$

(b) If $\alpha = 0.01$, the critical region is $z > 2.33$
Since 4.8 is 3 standard errors above 4.5, H_o is rejected.

8-21 H_o will be rejected if $(\bar{x} - 20.0)/0.5 > 1.65$

That is, H_o will be rejected if $\bar{x} > 20.825$

Therefore, any sample mean larger than 20.825 would cause z* to exceed 1.65, and would result in the rejection of H_o.

8-22 H_o: $\mu = 4.9$ $\alpha = 0.05$

H_a: $\mu > 4.9$

$z = \dfrac{5.3 - 4.9}{1.8/\sqrt{36}}$

$z^* = 1.33$

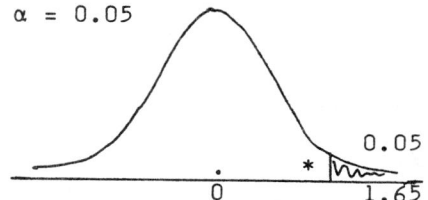

Fail to reject H_o, the sample does not provide sufficient evidence to reject the null hypothesis.

8-25 H_o: $\mu = 1600$ $\alpha = 0.02$

H_a: $\mu < 1600$ (less than)

$z = \dfrac{1562.3 - 1600}{150/\sqrt{100}}$

$z^* = -2.51$

Reject H_o, the sample does provide sufficient evidence to support the foreman's contentions, the mean is less than 1600.

8-28 (a) H_a: r>a   Failure to reject H_o will result in the drug being marketed. Because of the high current mortality rate, burden of proof is on the old ineffective drug.

(b) H_a: r<a   Failure to reject H_o will result in the new drug not being marketed. Because of the low current mortality rate, burden of proof should be on the new drug.

8-29 (a) P = P(z>1.48) = 0.5000 - 0.4306 = <u>0.0694</u>

(b) P = P(z<-0.85) = 0.5000 - 0.3023 = <u>0.1977</u>

(c) P = P(z<-1.17) + P(z>1.17) = 2(0.5000 - 0.3790)
      = <u>0.2420</u>

(d) P = P(z<-2.11) = 0.5000 - 0.4826 = <u>0.0174</u>

(e) P = P(z<-0.93) + P(z>0.93) = 2(0.5000 - 0.3238)
      = <u>0.3524</u>

(f) P = P(z>0.46) = 0.5000 - 0.1772 = <u>0.3228</u>

8-30 (a) $H_a: \mu > 35$ implies p = 0.0582 is area to right of z*

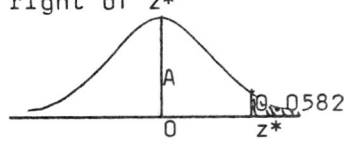

A = 0.4418, z* = <u>1.57</u>

(b) $H_a: \mu < 35$ implies p = 0.0166 is area to lefttt of z*

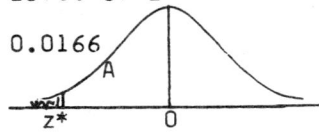

A = 0.4834, z* = <u>-2.13</u>

(c) $H_a: \mu \neq 35$ implies p = 0.0042 is total area for two tails

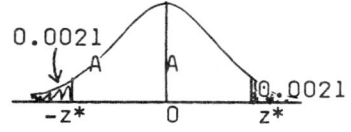

A = 0.4979, z* = 2.87

Therefore, <u>z* could be either -2.87 or +2.87.</u>

8-31 (a) Fail to reject H_o (b) Reject H_o

8-33 $H_o: \mu = 21.5$

$H_a: \mu > 21.5$ (increase)

$z = \dfrac{22.0 - 21.5}{2.5/\sqrt{64}}$

z* = 1.60

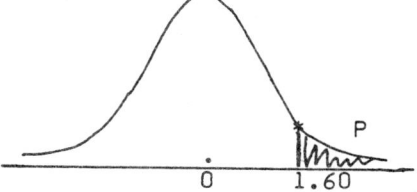

P = P(z>1.60) = 0.5000 - 0.4452 = <u>0.0548</u>

-52-

8-35   width = $(\bar{x} - z(\alpha/2)(\sigma/\sqrt{n})) - (\bar{x} - z(\alpha/2)(\sigma/\sqrt{n}))$
        = $2z(\alpha/2)\sigma/\sqrt{n}$

(a)   The higher the level of confidence, the larger the value of $z(\alpha/2)$, and the greater the width. Therefore, <u>as level of confidence increases, the width of the confidence interval increases.</u>

(b)   The larger the sample size n is, the smaller the standard error gets, and the smaller the width. Therefore, <u>as the sample size n increases, the width of the confidence interval decreases.</u>

(c)   The more variable the characteristic measured is, the larger the standard deviation will be, the larger the standard error will be, and the greater the width. Therefore, <u>as the standard deviation increases, the width of the confidence interval increases.</u>

8-37   (a) <u>7280.00</u>

(b) $7280 \pm 1.96\dfrac{1200}{\sqrt{36}}$

      7280 ± 392

      <u>6888 to 7672</u>, the 0.95 confidence interval for μ

(c) $7280 \pm 2.58\dfrac{1200}{\sqrt{36}}$

      7280 ± 516

      <u>6764 to 7796</u>, the 0.99 confidence interval for μ

8-39   (a) <u>25.3</u>

(b) $25.3 \pm 1.96\dfrac{4}{\sqrt{60}}$

      25.3 ± 1.01

      <u>24.29 to 26.31</u>, the 0.95 confidence interval for μ

(c) $25.3 \pm 2.58\dfrac{4}{\sqrt{60}}$

      25.3 ± 1.33

      <u>23.97 to 26.63</u>, the 0.99 confidence interval for μ

8-41 $n = \{\dfrac{(2.58)(900)}{75}\}^2 = 958.5 = \underline{959}$

8-45 (a) $H_0: \mu = 100$ (b) $H_a: \mu \neq 100$
(c) $\alpha = 0.01$ (d) $z(\alpha/2) = \pm 2.58$
(e) $\mu = 100$ (f) $\bar{x} = 96$ (g) $\sigma = 12$
(h) $\sigma_{\bar{x}} = \dfrac{12}{\sqrt{50}} = 1.697 = 1.7$
(i) $z^* = \dfrac{96-100}{1.7} = -2.35$
(j)   Fail to reject H_0
(k) $\alpha = 0.01$

```
              0.005              0.005
              ⌢*                 ⌢
            -2.58       0       2.58
```

8-46 (a) $\bar{x} = 32.0$ (b) $\sigma = 2.4$ (c) $n = 64$
(d) $1-\alpha = 0.90$ (e) $z(\alpha/2) = 1.65$
(f) $\sigma_{\bar{x}} = \dfrac{2.4}{\sqrt{64}} = 0.3$
(g) $E = (1.65)(0.3) = 0.495$
(h)   U.C.L. $= 32.0 + 0.495 = 32.495$
(i)   L.C.L. $= 32.0 - 0.495 = 31.505$

8-48 $H_0: \mu = 22$ $\alpha = 0.05$
      $H_a: \mu > 22$ (higher)

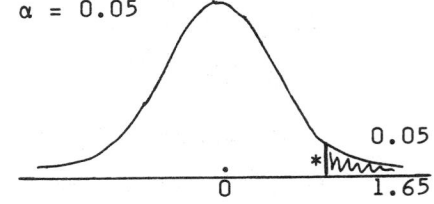

(a)   cut-off $\bar{x} = 22 + 1.65\dfrac{2.50}{\sqrt{45}} = \underline{22.61}$
(b) $\bar{x} = \dfrac{1010.25}{45} = 22.45$

No there is not sufficient evidence to reject the 22.00 claim of the admissions office (the null hypothesis).

8-51 (a) Classical approach:
$H_0: \mu = 9$ $\alpha = 0.02$

$H_a: \mu > 9$

$z = \dfrac{10.6-9.0}{2.5/\sqrt{24}}$

$z^* = 3.14$

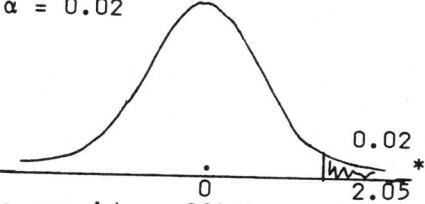

Reject H_0, the sample does provide sufficient evidence to conclude the mean waiting time is more than 9 minutes.

(b) Prob-value approach:
$H_0: \mu = 9$ $\alpha = 0.02$

$H_a: \mu > 9$

$z = \dfrac{10.6-9.0}{2.5/\sqrt{24}}$

$z = 3.14$

$P = P(z>3.14) = 0.5000 - 0.4992 = \underline{0.0008}$

Reject H_0, the sample does provide sufficient evidence to conclude the mean waiting time is more than 9 minutes.

8-55   The level of confidence determines the $z(\alpha/2)$ to be used. z is the number of multiples of the standard error used to determine the maximum error.

8-58   (a) $1.52 \pm 1.96\dfrac{\sqrt{2.25}}{\sqrt{20}}$

1.52 ± 0.66

$\underline{0.86 \text{ to } 2.18}$, the 0.95 confidence interval for μ

(b) $1.52 \pm 1.96\dfrac{\sqrt{2.25}}{\sqrt{32}}$

1.52 ± 0.52

$\underline{1.00 \text{ to } 2.04}$, the 0.95 confidence interval for μ

(c) The larger sample size causes a narrower interval.

8-60 (a) $95 \pm 1.28\dfrac{10}{\sqrt{25}}$

95 ± 2.56

$\underline{92.44 \text{ to } 97.56}$, the 0.80 confidence interval for μ

(b) $95 \pm 1.28\dfrac{10}{\sqrt{100}}$

95 ± 1.28

$\underline{93.72 \text{ to } 96.28}$, the 0.80 confidence interval for μ

(c) $95 \pm 1.28 \dfrac{5}{\sqrt{25}}$

95 ± 1.28

93.72 to 96.28, the 0.80 confidence interval for μ

8-62   (a) $142.00 \pm 1.96 \dfrac{70}{\sqrt{64}}$

142.00 ± 17.15

124.85 to 159.15, the 0.95 confidence interval for μ

(b)   2,497,000 to 3,183,000, the 0.95 confidence interval for Total.

8-64 $n = \{\dfrac{(2.58)(\sigma)}{\sigma/3}\}^2 = 59.9 = \underline{60}$

Chapter Nine

9-1 (a) 1.71 (b) 1.37 (c) 2.60 (d) 2.08
(e) -1.72 (f) -2.06 (g) -2.47 (h) 1.96

9-2 (a) t(19,0.05) = 1.73
(b) t(3,0.975) = -3.18, t(3,0.025) = 3.18
(c) t(18,0.99) = -2.55 (d) t(17,0.10) = 1.33
(e) t(7,0.95) = -1.89, t(7,0.05) = 1.89

9-3 df = 7

9-5 (a) 1 - (0.10 + 0.01) = 0.89
(b) 1 - (0.05 + 0.005) = 0.945

9-7 (a) Symmetric about mean, mean equals zero.
(b) Standard deviation of t-distribution is larger than one, t-distribution is different for each different sample size while there is only one z-distribution.

9-9 (a) Classical approach:
H_0: μ = 25 (at least) α = 0.01
H_a: μ < 25 (less than)
$t = \dfrac{19.4 - 25.0}{9.6/\sqrt{36}}$
$t^* = -3.50$

Reject H_0, the sample does provide sufficient evidence to reject the student's claim that the mean travel time is at least 25 minutes.

(b) Prob-value approach:
H_0: μ = 25 α = 0.01
H_a: μ < 25
$t = \dfrac{19.4 - 25.0}{9.6/\sqrt{36}}$
$t = -3.50$

$P = P(t < -3.50,$ with df=35$) = P(z < -3.50)$
$= 0.5000 - 0.4998 = 0.0002$

Reject H_0, the sample does provide sufficient evidence to reject the student's claim that the mean travel time is at least 25 minutes.

9-12 $12.6 \pm 1.71 \frac{3.0}{\sqrt{25}}$

12.60 ± 1.03

<u>11.57 to 13.63</u>, the 0.90 confidence interval for μ

9-14 $n = 10, \quad \Sigma x = 860, \quad \Sigma x^2 = 75222$

$\bar{x} = 860/10 = \underline{86.0}$

$s^2 = \frac{75222 - (860)^2/10}{9} = 140.222$

$s = \sqrt{140.222} = \underline{11.84}$

$86.00 \pm 1.83 \frac{11.84}{\sqrt{10}}$

86.00 ± 6.85

<u>79.15 to 92.85</u>, the 0.90 confidence interval for μ

9-16   Since x has a binomial distribution and $n = 14$, we may use table 6 to obtain these probabilities. The null hypothesis assumes that $p = 0.5$.

(a) $\alpha = P(x = 10,11,12,13,14 | B(n=14,p=0.5))$
$\alpha = 0.061 + 0.022 + 0.006 + 0.001 + 0.000 = \underline{0.090}$

(b) $\alpha = P(x = 11,12,13,14 | B(n=14,p=0.5))$
$\alpha = 0.022 + 0.006 + 0.001 + 0.000 = \underline{0.029}$

(c) $\alpha = P(x = 12,13,14 | B(n=14,p=0.5))$
$\alpha = 0.006 + 0.001 + 0.000 = \underline{0.007}$

9-18   (a) correctly fail to reject H_o

(b) $\alpha = P(x > 13.5 | \mu = 25/3$ and $\sigma = \sqrt{50/9})$

$= P(z > \frac{13.5 - 8.33}{2.357})$

$= P(z > 2.19)$
$= 0.5000 - 0.4857 = \underline{0.0143}$

(c)   commit a type II error

(d) $P = P(x > 12.5 | \mu = 25/3$ and $\sigma = \sqrt{50/9})$

$= P(z > \frac{12.5 - 8.33}{2.357})$

$= P(z > 1.77)$
$= 0.5000 - 0.4616 = \underline{0.0384}$

9-19   (a) Classical approach:
H$_0$: p = 0.90   α = 0.05

H$_a$: p < 0.90 (less than)

p' = x/n = 55/75 = 0.733

$z = \dfrac{0.733 - 0.900}{\sqrt{(0.90)(0.10)/75}}$

z* = -4.77

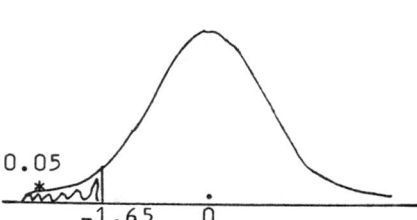

Reject H$_0$, the sample does provide sufficient evidence to show that less than 90% of the claims are settled within 30 days.

(b)   Prob-value approach:
H$_0$: p = 0.90

H$_a$: p < 0.90

p' = x/n = 55/75 = 0.733

$z = \dfrac{0.733 - 0.900}{\sqrt{(0.90)(0.10)/75}}$

z = -4.77

P = P(z<-4.77) < 0.000003

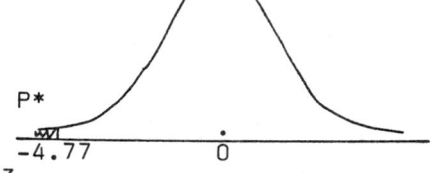

We should doubt the insurance company's claim.

9-23   (a) $0.3 \pm 1.65\sqrt{\dfrac{(0.3)(0.7)}{30}}$

= 0.3 ± (1.65)(0.0837) = 0.300 ± 0.138

lower limit = 0.162,   upper limit = 0.438

(b) $0.7 \pm 1.65\sqrt{\dfrac{(0.7)(0.3)}{30}}$

= 0.7 ± (1.65)(0.0837) = 0.700 ± 0.138

lower limit = 0.562,   upper limit = 0.838

(c) $0.5 \pm 1.65\sqrt{\dfrac{(0.5)(0.5)}{10}}$

= 0.5 ± (1.65)(0.158) = 0.500 ± 0.26

lower limit = 0.24,   upper limit = 0.76

(d) $0.5 \pm 1.65\sqrt{\dfrac{(0.5)(0.5)}{100}}$

= 0.5 ± (1.65)(0.158) = 0.500 ± 0.0825

lower limit = 0.418,   upper limit = 0.582

(e) $0.5 \pm 1.65\sqrt{\frac{(0.5)(0.5)}{1000}}$

$= 0.5 \pm (1.65)(0.0158) = 0.500 \pm 0.026$

lower limit = $\underline{0.474}$, upper limit = $\underline{0.526}$

(f) The two answers are symmetric about 0.5.

(g) As the sample size increased the interval became narrower

9-25 $E = z(\alpha/2)\sqrt{\frac{pq}{n}}$

(a) $E = 1.96\sqrt{(0.1)(0.9)/1000} = 1.96(0.0094) = \underline{0.019}$

(b) $E = 1.96\sqrt{(0.2)(0.8)/100} = 1.96(0.040) = \underline{0.078}$

(c) $E = 1.96\sqrt{(0.5)(0.5)/50} = 1.96(0.071) = \underline{0.139}$

9-27   (a) $p' = 75/350 = \underline{0.21}$

(b) $E = 1.96\sqrt{(0.21)(0.79)/350} = 1.96(0.022) = \underline{0.043}$

We are 95% confident that our estimate, 0.21, is no more than 0.05 different than the true proportion, p.

9-30   Let p = P(used at least one other service in last 6 months) Use formula (9-6):

$n = \frac{(2.33)^2(0.4)(0.6)}{(0.05)^2} = 521.17 = \underline{522}$

9-33   (a) 34.8   (b) 28.9   (c) 13.4   (d) 48.3

(e) 12.3   (f) 3.25   (g) 37.7   (h) 10.9

9-34   (a) $\chi^2(19, 0.05) = 30.1$

(b) $\chi^2(4, 0.01) = 13.3$

(c) $\chi^2(17, 0.975) = 7.56$

(d) $\chi^2(60, 0.95) = 43.2$

(e) $\chi^2(21, 0.95) = 11.6$ $\chi^2(21, 0.05) = 32.7$

(f) $\chi^2(6, 0.975) = 1.24$ $\chi^2(6, 0.025) = 14.5$

(g) $\chi^2(7, 0.99) = 1.24$

(h) $\chi^2(17, 0.99) = 6.41$ $\chi^2(17, 0.01) = 33.4$

9-37   1 - (0.01 + 0.05) = 0.94

9-38   (a) Classical approach:
        H_0: $\sigma = 0.25$(no increase) $\alpha = 0.10$

        H_a: $\sigma > 0.25$(increase)

        $\chi^2 = \frac{(20-1)(0.35)^2}{(0.25)^2}$

        $\chi^{2*} = 37.24$

        0.10

        0   19   27.2

        Reject H_0, the sample does provide sufficient
        evidence to show an increase.

(b)   Prob-value approach:
        H_0: $\sigma = 0.25$(no increase) $\alpha = 0.10$

        H_a: $\sigma > 0.25$(increase)

        $\chi^2 = \frac{(20-1)(0.35)^2}{(0.25)^2}$

        $\chi^{2*} = 37.24$

        0   19   37.24

        $P = P(\chi^2$ with df $= 19 > 37.24)$, $0.005 < P < 0.010$

        Reject H_0, the sample does provide sufficient
        evidence to show an increase.

9-41   Summary of data:
        n = 10 $\Sigma x = 5493$ $\Sigma x^2 = 3,034,605$

        $SS(x) = 3,034,605 - 5493^2/10 = 17,300.1$
        $s = \sqrt{17,300.1/9} = \sqrt{1922.233} = 43.8433$

        $43.8433\sqrt{\frac{9}{19.0}}$ to $43.8433\sqrt{\frac{9}{2.70}}$

        (30.2 to 80.8), 0.95 confidence interval for σ

9-43   (a) Classical approach:
        H_0: $\mu = 80$(same as state) $\alpha = 0.05$

        H_a: $\mu \neq 80$(different)

        $t = \frac{77.5-80.0}{2.5/\sqrt{20}}$

        $t^* = -4.47$ *

        0.025   0.025

        -2.09   0   2.09

        Reject H_0, the sample does provide sufficient
        evidence to show the mean is different than 80.

(b) Prob-value approach:
$H_0: \mu = 80$
$H_a: \mu \ne 80$
$t = \dfrac{77.5-80.0}{2.5/\sqrt{20}}$
$t = -4.47$

$\alpha = 0.05$

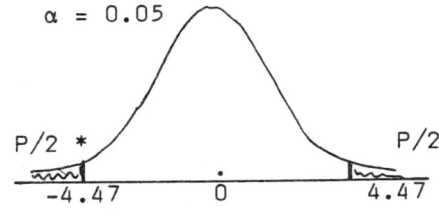

$P = 2P(t>4.47) < \underline{0.001}$

Reject H_0, there does seem to be a difference

9-46 Summary of data:
$n = 21$, $\Sigma x = 241.2$, $\Sigma x^2 = 2774.86$

(a) $\bar{x} = 241.2/21 = \underline{11.49}$

$s^2 = \dfrac{2774.86 - (241.2)^2/21}{20} = 0.2252$

$s = \sqrt{0.2252} = \underline{0.47}$

(b) $11.49 \pm 2.53 \dfrac{0.47}{\sqrt{21}}$

11.49 ± 0.26

$\underline{11.23 \text{ to } 11.75}$, the 0.98 confidence interval for μ

9-48 (a) $P = P(t < -2.01)$, $\underline{0.025 < P < 0.05}$

(b) $P = P(t > 2.01)$, $\underline{0.025 < P < 0.05}$

(c) $P = P(t < -2.01) + P(t > +2.01)$, $\underline{0.05 < P < 0.10}$

(d) $P = P(t < -2.01) + P(t > +2.01)$, $\underline{0.05 < P < 0.10}$

9-51 (a) $\bar{x} = 878.2/100 = \underline{8.782}$

$s = \sqrt{49.91/99} = \underline{0.71}$

(b) point estimate for $\mu = \bar{x} = \underline{8.782}$

(c) $8.782 \pm 1.96 \dfrac{0.71}{\sqrt{100}}$

8.782 ± 0.139

$\underline{8.643 \text{ to } 8.921}$, the 0.95 confidence interval for μ

(d) point estimate for $\sigma = s = \underline{0.71}$

(e) $\sqrt{\dfrac{(99)(0.504)}{128.87}}$ to $\sqrt{\dfrac{(99)(0.504)}{73.3}}$

$\sqrt{0.3874}$ to $\sqrt{0.6807}$

$\underline{0.622 \text{ to } 0.825}$, the 0.95 interval for σ

9-54   Let x = number of defectives in the sample of 50.
       When H_o is true, we may treat x as a binomial
       variable with n = 50 and p = 0.005.

   α is the probability of rejecting H_o when it is true, or
   $\alpha = P(x > 2 | B(n = 50, p = 0.005))$
   $\alpha = 1 - P(x = 0,1 | B(n=50, p=0.005))$
   $= 1 - \{\binom{50}{0}(0.005)^0(0.995)^{50}\} + \{\binom{50}{1}(0.005)^1(0.995)^{49}\}$
   $= 1 - (0.7783 + 0.1956)$
   $\alpha = \underline{0.0261}$

9-57 $p' = 250/450 = 0.556$

   $0.556 \pm 1.96\sqrt{\frac{(0.556)(0.444)}{450}}$

   0.556 ± 0.023

   $\underline{0.533 \text{ to } 0.579}$, the 0.95 interval for p

9-60   (a) $0.233 \pm 1.96\sqrt{\frac{(0.233)(0.767)}{60}}$

   0.233 ± 0.107

   $\underline{0.126 \text{ to } 0.340}$, the 0.95 interval for p, P(dissat.)

(b)   The dealer has overestimated his percent of satisfied customers, it appears to be less than 90 %.

9-63   With no estimate for p, use p = q = 0.5.

   $n = \frac{(1.96)^2(0.5)(0.5)}{(0.01)^2} = \underline{9604}$

9-67 χ^{2*} would need to exceed $\chi^2(29, 0.05) = 42.6$

   $\frac{(29)(s^2)}{17} > 42.6$

   $\underline{s^2 > 25}$

9-69   (a) Summary: n = 12, $\Sigma x = 41.3$, $\Sigma x^2 = 146.83$

   $\bar{x} = 41.3/12 = 3.44$

   $s^2 = \frac{146.83 - (41.3)^2/12}{11}$

   $s^2 = 0.4263$

   $s = \sqrt{0.4263} = 0.653$

(b) $H_o: \mu = 4.9$ $\alpha = 0.05$

$H_a: \mu < 4.9$

$t = \dfrac{3.44 - 4.90}{0.653/\sqrt{12}}$

$t^* = -7.75$

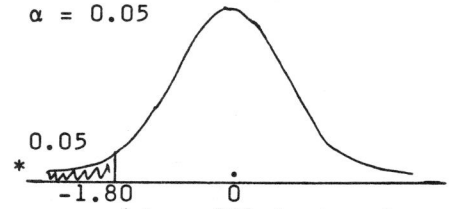

Reject H_o, the sample does provide sufficient evidence to claim that the mean carbon monoxide is low.

(c) $H_o: \sigma^2 = 0.25$ (no more than) $\alpha = 0.05$

$H_a: \sigma^2 > 0.25$ (more than)

$\chi^2 = \dfrac{(12-1)(0.4263)}{(0.25)}$

$\chi^{2*} = 18.8$

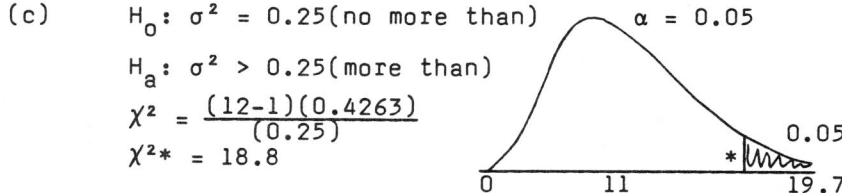

Fail to reject H_o, the sample does not provide sufficient evidence to reject H_o.

9-72   Formula (9-5): $E = z(\alpha/2)\sqrt{\dfrac{pq}{n}}$

$E = \dfrac{z(\alpha/2)\sqrt{pq}}{\sqrt{n}}$

$E\sqrt{n} = z(\alpha/2)\sqrt{pq}$

$\sqrt{n} = \dfrac{z(\alpha/2)\sqrt{pq}}{E}$

$n = \{\dfrac{z(\alpha/2)\sqrt{pq}}{E}\}^2$

Formula (9-6): $n = \dfrac{\{z(\alpha/2)\}^2 pq}{E^2}$

Chapter Ten

10-1 Dependent samples. The sources used for both samples are the same.

10-4 Independent samples. The two samples are from two unrelated sets of rats.

10-7 (a) $P(|\bar{x}_1 - \bar{x}_2| > 0.002)$

$$= P\left(|z| > \frac{0.002 - 0.000}{\sqrt{\frac{0.0004}{100} + \frac{0.0004}{80}}}\right)$$

$= P(|z| > 0.67) = 2(0.5000 - 0.2486) = \underline{0.5028}$

(b) $P(|\bar{x}_1 - \bar{x}_2| < 0.0015)$

$$= P\left(|z| < \frac{0.0015 - 0.000}{\sqrt{\frac{0.0004}{100} + \frac{0.0004}{80}}}\right)$$

$= P(|z| < 0.50) = 2(0.1915) = \underline{0.3830}$

10-8 (a) $P = P(z > 1.85) = 0.5000 - 0.4678 = \underline{0.0322}$

(b) $P = P(z < -2.33) + P(z > 2.33)$
$= 2(0.5000 - 0.4901) = 2(0.0099) = \underline{0.0198}$

(c) $P = P(z < -2.76) = 0.5000 - 0.4971 = \underline{0.0029}$

10-10 (a) Classical approach:

$H_0: \mu_I - \mu_{II} = 0$

$H_a: \mu_I - \mu_{II} \neq 0$

$\alpha = 0.10$

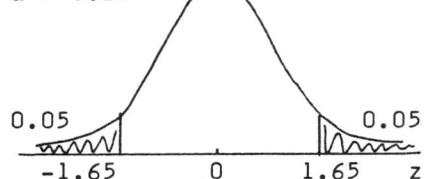

0.05   0.05
-1.65   0   1.65   z

Using formula (10-1):

$$z = \frac{(104.5 - 110.9) - 0}{\sqrt{\frac{200}{32} + \frac{700}{40}}}$$

$z^* = -1.31$

Fail to reject H_0, we are not able to reject the null hypothesis, the two population means are equal.

-65-

(b)   Prob-value approach:
$H_o: \mu_A - \mu_B = 0$

$H_a: \mu_A - \mu_B \neq 0$

$\alpha = 0.05$

Using formula (10-1):

$$z = \frac{(104.5 - 110.9) - 0}{\sqrt{\frac{200}{32} + \frac{700}{40}}}$$

$z^* = -1.31$

$P = P(|z| > 1.31) = 2(0.5000 - 0.4049) = \underline{0.1902}$

10-13 (a) Classical approach:
$H_o: \mu_c - \mu_e = 0$

$H_a: \mu_c - \mu_e = < 0$

$\alpha = 0.05$

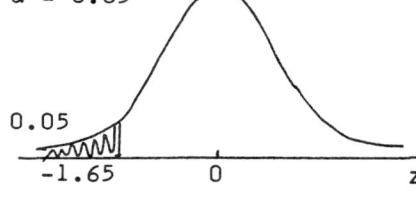

Using formula (10-3):

$$z = \frac{(5.2 - 7.6) - 0}{\sqrt{\frac{1.1^2}{50} + \frac{1.3^2}{50}}}$$

$z^* = -10.0$

Reject H_o, we have sufficient evidence to reject the null hypothesis, that is, the experimental group of rats consumed a significantly larger amount of lead.

(b)   Prob-value approach:
$H_o: \mu_c - \mu_e = 0$

$H_a: \mu_c - \mu_e = < 0$

$\alpha = 0.05$

Using formula (10-3):

$$z = \frac{(5.2 - 7.6) - 0}{\sqrt{\frac{1.1^2}{50} + \frac{1.3^2}{50}}}$$

$z^* = -10.0$

$P = P(z < -10.00) < 0.000001$

Reject H_o, we have sufficient evidence to reject the null hypothesis, that is, the experimental group of rats consumed a significantly larger amount of lead.

10-15 $(15.5 - 14.3) \pm 1.65\sqrt{\frac{2.7^2}{125} + \frac{3.0^2}{115}}$

1.2 ± 0.6

<u>0.6 to 1.8</u>, the 0.90 interval for $\mu_1 - \mu_2$

10-17 (a) F(15,17,0.99) and F(15,17,0.01)
(b) F(7,19,0.99)   (c) F(7,19,0.05)
(d) F(19,24,0.975) and F(19,24,0.025)
(e) F(24,34,0.95)

10-18 (a) 2.51   (b) 2.20   (c) 2.91   (d) 4.10
(e) 2.67   (f) 3.77   (g) 1.79   (h) 2.99

10-22 $H_o: \sigma_k^2 = \sigma_m^2$
$H_a: \sigma_k^2 \neq \sigma_m^2$

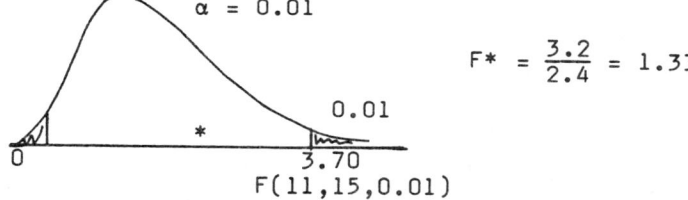

$\alpha = 0.01$

$F^* = \frac{3.2}{2.4} = 1.33$

Fail to reject H_o, there is not sufficient evidence to reject the null hyothesis.

(b) $P = P(F>1.33) > 0.10$

10-24 $(1.25/1.18)/F(24,24,0.025)$ to $(1.25/1.18) \times F(24,24,0.025)$

$1.06/2.27$ to 1.06×2.27

<u>0.47 to 2.41</u>, the 95% confidence interval for σ_1^2/σ_2^2

10-27 (a) Formula (10-11) is used when it is assumed that $\sigma_1 = \sigma_2$, and formula (10-12) is used when it is assumed that the standard deviations are not equal. Notice that a different formula is used to estimate the standard error.

(b) The critical region become larger in the case where it is assumed that the standard deviations are not equal.

(c) The number of degrees of freedom is smaller in the case where it is assumed that the standard deviations are not equal.

10-28   Sample information:

R: $\bar{x} = 295/10 = 29.5$, $s = \sqrt{75/9} = 2.89$

S: $\bar{x} = 195/8 = 24.4$, $s = \sqrt{90/7} = 3.59$

1st: H_o: $\sigma_R^2 = \sigma_S^2$

H_a: $\sigma_R^2 \neq \sigma_S^2$

$\alpha = 0.05$

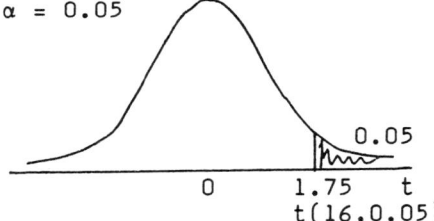

$F* = \dfrac{12.86}{8.33} = 1.54$

0.025

4.20
F(7,9,0.025)

Fail to reject H_o, therefore assume $\sigma_R^2 = \sigma_S^2$ and use Case I methods.

2nd: H_o: $\mu_R - \mu_S = 0$

H_a: $\mu_R - \mu_S > 0$

$\alpha = 0.05$

0.05

0   1.75   t
t(16,0.05)

Using formula (10-11):

$t = \dfrac{(29.5 - 24.4) - (0)}{\sqrt{\dfrac{(75 + 90)}{8+10-2}}\sqrt{\dfrac{1}{8} + \dfrac{1}{10}}}$

$t* = 3.348$

Reject H_o, we have sufficient evidence to show that μ_R is larger than μ_S.

10-31   P = 0.394 means that if the null hypothesis is rejected, there is a high probability of committing the type I error.

-68-

10-32 (a) Classical approach:

1st: $F(17,17,0.025) = 2.68$ and $F^* = 2.0$,

therefore assume $\sigma_A = \sigma_B$ and use Case 1 methods.

2nd: $H_0: \mu_S - \mu_n = 0$

$H_a: \mu_S - \mu_n \neq 0$

$\alpha = 0.05$

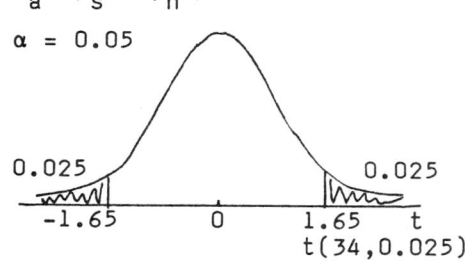

0.025   0.025
-1.65   0   1.65   t
              t(34,0.025)

Using formula (10-11):

$$t = \frac{(15{,}000 - 16{,}000) - (0)}{\sqrt{\frac{(17)(2400)+(17)(4800)}{18+18-2}}\sqrt{\frac{1}{18}+\frac{1}{18}}}$$

$t^* = -50.00$

Reject H_0, we have sufficient evidence to show that μ_S is not equal to μ_n.

(b) Prob-value approach:

1st: $F(17,17,0.025) = 2.68$ and $F^* = 2.0$,
therefore assume $\sigma_A = \sigma_B$ and use Case 1 methods.

2nd: $H_0: \mu_S - \mu_n = 0$

$H_a: \mu_S - \mu_n \neq 0$

$\alpha = 0.05$

Using formula (10-11):

$$t = \frac{(15{,}000 - 16{,}000) - (0)}{\sqrt{\frac{(17)(2400)+(17)(4800)}{18+18-2}}\sqrt{\frac{1}{18}+\frac{1}{18}}}$$

$t^* = -50.00$

$P = P(|t| > 50.00) = 0.0+$

Reject H_0, we have sufficient evidence to show that μ_S is not equal to μ_n.

10-34(a) $H_0: \sigma_1^2 = \sigma_2^2$

$H_a: \sigma_1^2 \neq \sigma_2^2$

$\alpha = 0.05$

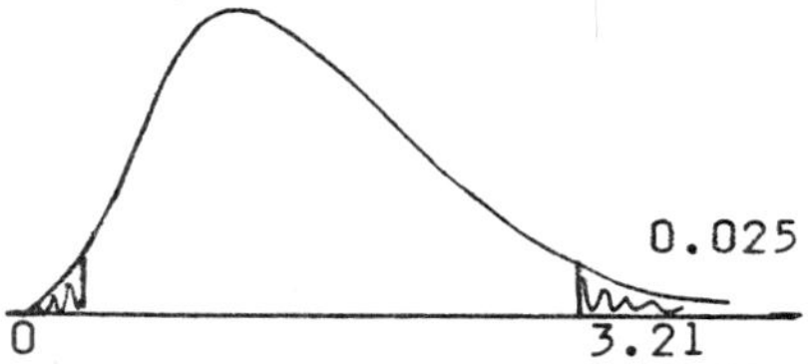

$F* = \dfrac{7.5^2}{2.1^2} = 12.8$

0.025

3.21
F(9,14,0.025)

Reject H_0, therefore assume $\sigma_1^2 \neq \sigma_2^2$ and use Case 2 methods.

(b) $H_0: \mu_2 - \mu_1 = 0$

$H_a: \mu_2 - \mu_1 > 0$

$\alpha = 0.05$

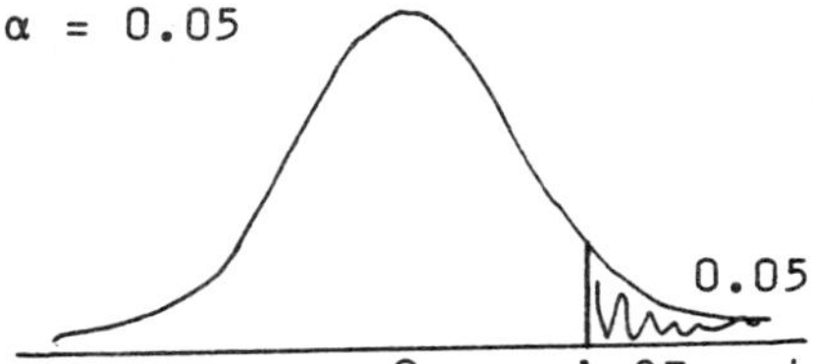

0.05

1.83 t
t(9,0.05)

Using formula (10-12):

$$t = \dfrac{(67.2 - 60.3) - (0)}{\sqrt{\dfrac{2.1^2}{15} + \dfrac{7.5^2}{10}}}$$

$t* = 2.84$

Reject H_0, we have sufficient evidence to show that the mean score for those with computer experience is significantly less than the mean score for those without the computer experience.

10-37 $(29.50 - 24.375) \pm 2.12 \sqrt{\dfrac{75 + 90}{10 + 8 - 2}} \sqrt{\dfrac{1}{10} + \dfrac{1}{8}}$

$5.125 \pm 2.12\sqrt{10.3125}\sqrt{0.225}$

5.125 ± 3.227

1.90 to 8.35, the 0.95 interval for $\mu_R - \mu_S$

10-40 $s_p\sqrt{\dfrac{1}{n_1}+\dfrac{1}{n_2}} = \sqrt{\dfrac{(n-1)s_1^2+(n-1)s_2^2}{n+n-2}}\sqrt{\dfrac{1}{n}+\dfrac{1}{n}}$

$= \sqrt{\dfrac{(n-1)(s_1^2+s_2^2)}{2(n-1)}}\sqrt{\dfrac{2}{n}}$

$= \sqrt{\dfrac{2(n-1)(s_1^2+s_2^2)}{2(n-1)(n)}}$

$= \sqrt{\dfrac{s_1^2+s_2^2}{n}}$

10-42 Sample statistics: $\bar{d} = 5.5$, $s_d = 11.34$

(a) Classical approach:
$H_o: \mu_d = 0$ $\alpha = 0.01$
$H_a: \mu_d > 0$ (beneficial)
$t = \dfrac{5.5 - 0.0}{11.34/\sqrt{40}}$
$t^* = 3.06$

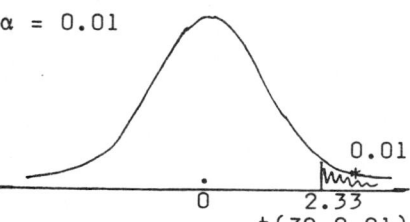

Reject H_o, there is a significant benefit to coating the pipe.

(b) Prob-value approach:
$H_o: \mu_d = 0$
$H_a: \mu_d > 0$ (beneficial) $\alpha = 0.05$
$t = \dfrac{5.5 - 0.0}{11.34/\sqrt{40}}$
$t^* = 3.06$

$P = P(t > 3.06) = P(z > 3.06) = 0.5000 - 0.4989$
$P = 0.0011$

Reject H_o, there is a significant benefit to coating the pipe.

10-45 d = 7 -2 9 2 2 10 -6 8, $\Sigma d = 30$ $\Sigma d^2 = 342$

$\bar{d} = 3.75$ $s_d = 5.726$ $t(7, 0.025) = 2.36$

$3.75 \pm 2.36(5.726/\sqrt{7}) = 3.75 \pm 5.10$

-1.35 to 8.85, the 95% confidence interval for μ_d

10-47 (a) Classical approach:
$H_o: p_A - p_B = 0$
$H_a: p_A - p_B \neq 0$
$\alpha = 0.02$

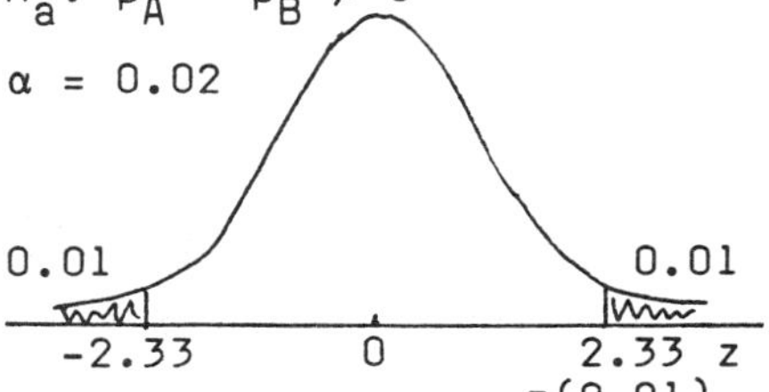

0.01 ⟍ ⟋ 0.01
-2.33   0   2.33 z
            z(0.01)

$$z = \frac{\frac{102}{300} - \frac{152}{400}}{\sqrt{(\frac{102+152}{300+400})(1-\frac{254}{700})(\frac{1}{300}+\frac{1}{400})}}$$

$$= \frac{0.34 - 0.38}{\sqrt{(0.363)(0.637)(0.005833)}}$$

$z^* = -1.09$

Fail to reject H_o, there is not sufficient evidence to show a difference.

(b) Prob-value approach:
The first several steps can be found in part (a).

$P = P(|z| > 1.09) = 2(0.5000 - 0.3621) = \underline{0.2758}$

Fail to reject H_o, there is not sufficient evidence to show a difference.

10-49 (a) Classical approach:
$H_o: p_c - p_m = 0$
$H_a: p_c - p_m \neq 0$
$\alpha = 0.05$

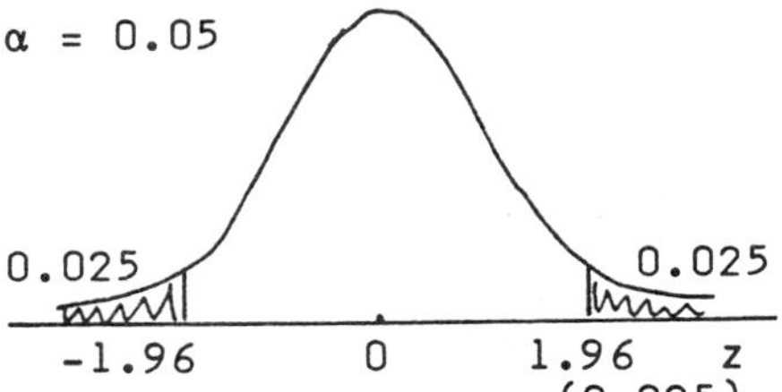

0.025 ⟍ ⟋ 0.025
-1.96   0   1.96 z
            z(0.025)

$$z = \frac{0.40 - 0.50}{\sqrt{(0.45)(0.55)(0.0200)}}$$

$z^* = -1.42$

Fail to reject H_o, there is not sufficient evidence to show a difference.

(b) Prob-value approach:
The first several steps can be found in part (a).

$P = P(|z| > 1.42) = 2(0.5000 - 0.4222) = \underline{0.1556}$

Fail to reject H_o, there is not sufficient evidence to show a difference.

10-51 (a) $E = 1.96\sqrt{(0.2)(0.8)(1/750 + 1/750)} = \underline{0.0405}$

Table B shows <u>5</u>, the calculated value rounded up.

(b) $E = 1.96\sqrt{(0.5)(0.5)(1/750 + 1/750)} = \underline{0.0506}$

Table C shows <u>6</u>, the calculated value rounded up.

(c) A decrease of 14% is significant. Anything 6% or over, the value from (b), is significant.

(d) A decrease of 2% is not significant. A change of anything less than 6% is not significant.

10-52 Use formula (10-21):

$(\frac{26}{40} - \frac{22}{40}) \pm z(0.04) \sqrt{\frac{(\frac{26}{40})(\frac{14}{40})}{40} + \frac{(\frac{22}{40})(\frac{18}{40})}{40}}$

$0.10 \pm 1.75\sqrt{0.011875}$
0.100 ± 0.191

$\underline{-0.091 \text{ to } 0.291}$, the 0.92 interval for $p_{bl} - p_{br}$

10-55 (a) Classical approach:
$H_o: \mu_t - \mu_1 = 0$
$H_a: \mu_t - \mu_1 \neq 0$
$\alpha = 0.05$

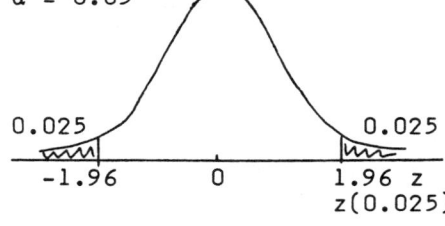

0.025   0.025
-1.96   0   1.96 z
              z(0.025)   (continued)

Using formula (10-1):

$z = \dfrac{(41.9 - 44.4) - 0}{\sqrt{\dfrac{12^2}{150} + \dfrac{12^2}{150}}}$

$z^* = -1.80$

Fail to reject H_o, there is no significant difference.

(b) Prob-value approach:
The first several steps can be found in part (a).

$$P = P(|z| > 1.80) = 2(0.5000 - 0.4641) = \underline{0.0718}$$

10-58 Using formula (10-3):

$$z = \frac{(70.5 - 75.7) - 0}{\sqrt{\frac{13.2^2}{50} + \frac{13.6^2}{50}}}$$

$z^* = -1.94$

$$P = P(z < -1.94) + P(z > 1.94)$$
$$= 2P(z > 1.94) = 2(0.5000 - 0.4738) = \underline{0.0524}$$

10-59 Use formula (10-2):

$$(73.2-70.5) \pm 1.65\sqrt{\frac{6.1^2}{70} + \frac{5.5^2}{60}}$$

2.70 ± 1.68

$\underline{1.02 \text{ to } 4.38}$, the 0.90 interval for $\mu_1 - \mu_2$

10-60 $H_o: \sigma_2^2 = \sigma_1^2$
$H_a: \sigma_2^2 \neq \sigma_1^2$
$\alpha = 0.05$

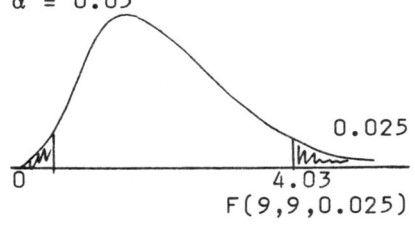

$F^* = \frac{17.3}{10.5} = 1.65$

Fail to reject H_o, she cannot conclude that her consistency (as measured by σ^2) is different on the two courses.

10-63 (a) $\frac{1.968}{2.834} = \underline{0.6944}$

(b) $0.6944/2.44$ to $0.6944(2.70)$

$\underline{0.2846 \text{ to } 1.8749}$, the 0.95 interval for σ_L^2/σ_H^2

(c) $\underline{0.533 \text{ to } 1.369}$, the 0.95 interval for σ_L/σ_H

10-65   Sample statistics.
$n_R = 7$, $\bar{x}_R = 7.29$, $s_R = 2.29$
$n_P = 8$, $\bar{x}_P = 2.25$, $s_P = 1.83$

(a) Classical approach:
1st: $H_0: \sigma_R^2 = \sigma_P^2$ $H_a: \sigma_R^2 \neq \sigma_P^2$
$\alpha = 0.10$ $F(6,7,0.05) = 3.87$
$F^* = 2.29^2/1.83^2 = 1.57$

Fail to reject H_0, therefore assume $\sigma_L^2 = \sigma_H^2$ and use Case 1 methods.

2nd: $H_0: \mu_R - \mu_P = 0$
$H_a: \mu_R - \mu_P \neq 0$
$\alpha = 0.10$

0.05   0.05
-1.77   0   1.77 t
                    t(13,0.05)

Using formula (10-9):
$$s_p = \sqrt{\frac{(7)(2.29)^2 + (8)(1.83)^2}{7 + 8 - 2}}$$
$s_p = \sqrt{4.8846} = 2.21$

Using formula (10-11):
$$t = \frac{(7.29 - 2.25) - (0)}{2.21\sqrt{\frac{1}{7} + \frac{1}{8}}}$$
$t^* = 4.41$

Reject H_0, there is a significant difference in the mean scores.

(b)   Prob-value approach:
The first several steps can be found in part (a).

$P = P(|t| > 4.68$, with df $= 13)$, $P < 0.01$

10-66 (a) $H_0: \sigma_A^2 = \sigma_B^2$

$H_a: \sigma_A^2 \neq \sigma_B^2$

$\alpha = 0.10$

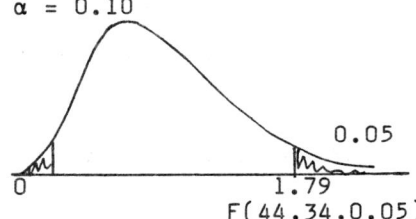

$F^* = \dfrac{7.00}{3.25} = 2.15$

Reject H_0, therefore assume $\sigma_L^2 \neq \sigma_H^2$ and use Case 2 methods.

(b) $H_0: \mu_A - \mu_B = 0$

$H_a: \mu_A - \mu_B > 0$

$\alpha = 0.10$

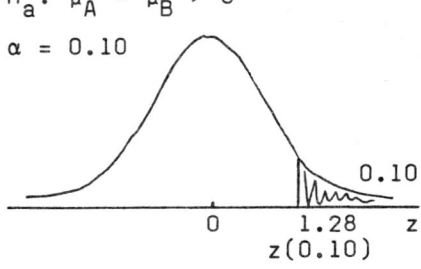

Using formula (10-13):

$$z = \dfrac{(10.3 - 7.3) - 0}{\sqrt{\dfrac{7.0}{45} + \dfrac{3.25}{35}}}$$

$t^* = 6.02$

Reject H_0, there is more weight loss with diet A.

(c) $(10.3 - 7.3) \pm 1.65 \sqrt{\dfrac{7.0}{45} + \dfrac{3.25}{35}}$

3.00 ± 0.82

<u>2.18 to 3.82</u>, the 0.90 interval for $\mu_A - \mu_B$

10-67   Less accurate would be implied if the mean target error of one is significantly greater than the mean target error of the other.

1st: $F(9,7,0.025) = 4.82$ and $F* = 1.44$,

therefore assume $\sigma_1 = \sigma_2$ and use Case 1 methods.

(a) $H_o: \mu_2 - \mu_1 = 0$

$H_a: \mu_2 - \mu_1 > 0$ (2nd type is less accurate)

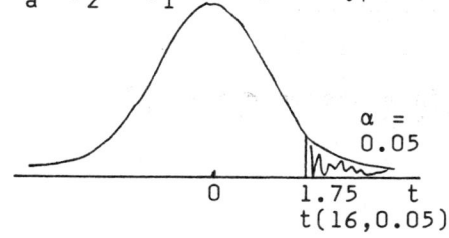

$\alpha = 0.05$

0   1.75   t
    $t(16, 0.05)$

Using formula (10-9):

$$S_p = \sqrt{\frac{(7)(15)^2 + (9)(18)^2}{8 + 10 - 2}}$$

$S_p = \sqrt{208.69} = 16.75$

Using formula (10-11):

$$t = \frac{(52 - 36) - (0)}{16.75\sqrt{\frac{1}{8} + \frac{1}{10}}}$$

$t* = 2.01$

Reject H_o, the second kind of rocket appears to be less accurate.

10-70   Summary of samples:

	n	Σx	Σx^2	$\bar{x}$	s^2
male	18	609	20,713	33.83	6.38
female	11	289	7,633	26.27	4.02

1st: $F(17,10,0.025) = 3.56$ and $F* = 6.38/4.02 = 1.59$,

therefore assume $\sigma_1 = \sigma_2$ and use Case 1 methods.

(a) $H_o: \mu_m - \mu_f = 0$

$H_a: \mu_m - \mu_f > 0$ (males larger)

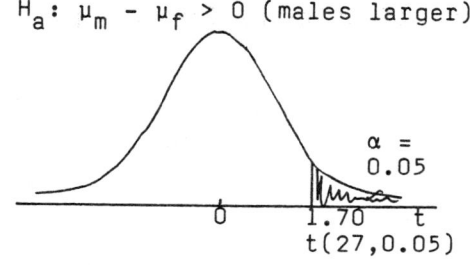

$\alpha = 0.05$

0   1.70   t
    $t(27, 0.05)$

continued

-77-

Using formula (10-9):

$$s_p = \sqrt{\frac{(17)(6.38) + (10)(4.02)}{18 + 11 - 2}}$$

$$s_p = \sqrt{5.5059} = 2.35$$

Using formula (10-11):

$$t = \frac{(33.83 - 26.27) - (0)}{2.35\sqrt{\frac{1}{18} + \frac{1}{11}}}$$

$t^* = 8.4$

Reject H_o, the mean waist size for men is larger.

10-74 Sample statistics: $d = x - y$, $n = 12$

$\bar{d} = 0.28$, $s_d = 0.43$

(a) Classical approach:
$H_o: \mu_d = 0$ $\alpha = 0.10$
$H_a: \mu_d \neq 0$ (is diff.)

$$t = \frac{0.28 - 0.0}{0.43/\sqrt{12}}$$

$t^* = 2.26$

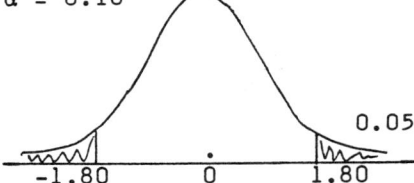

Reject H_o, there is evidence of a significant difference between the two mixtures.

(b) Prob-value approach:
The first several steps can be found in part (a).

$P = P(|t| > 2.26$, with df $= 11)$, $0.02 < P < 0.05$

10-76 Sample statistics: $d = $ after $-$ before, $n = 10$

$\bar{d} = 7.0$, $s_d = 5.8$

(a) Classical approach:
$H_o: \mu_d = 0$ $\alpha = 0.01$
$H_a: \mu_d > 0$ (improvement)

$$t = \frac{7.0 - 0.0}{5.8/\sqrt{10}}$$

$t^* = 3.82$

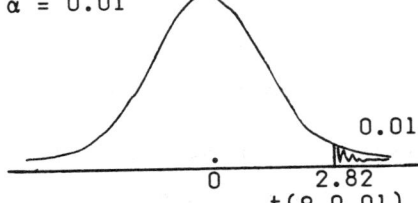

Reject H_o, there is evidence that a significant improvement has taken place.

(b) Prob-value approach:
The first several steps can be found in part (a).

$P = P(t > 3.82, \text{ with df} = 9) < 0.005$

10-78 $d = 0 - Y$, $n = 12$, $\bar{d} = 3.583$, $s_d = 19.58$

$3.583 \pm 2.20(19.58/\sqrt{12})$

3.583 ± 12.435

$\underline{-8.85 \text{ to } 16.02}$, the 0.95 interval for μ_d

10-80 (a) Classical approach:

$H_o: p_A - p_B = 0$
$H_a: p_A - p_B \neq 0$
$\alpha = 0.10$

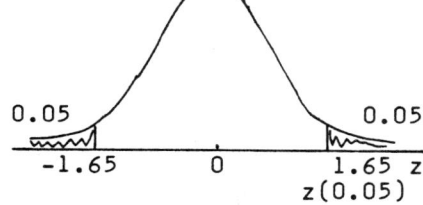

0.05   0.05
-1.65   0   1.65 z
              z(0.05)

$z = \dfrac{\dfrac{15}{250} - \dfrac{30}{300}}{\sqrt{(\dfrac{15+30}{250+300})(1-\dfrac{45}{550})(\dfrac{1}{250}+\dfrac{1}{300})}}$

$= \dfrac{0.06 - 0.10}{\sqrt{(0.082)(0.918)(0.0073)}}$

$z* = -1.71$

Reject H_o, there is a significant difference.

(b) Prob-value approach:
The first several steps can be found in part (a).

$P = P(|z| > 1.71) = 2(0.5000 - 0.4564) = \underline{0.0872}$

Reject H_o, there is a significant difference.

10-83 Use formula (10-21):

$(0.32 - 0.27) \pm 2.58\sqrt{\dfrac{(0.32)(0.68)}{100} + \dfrac{(0.27)(0.73)}{100}}$

0.050 ± 0.166

$\underline{-0.116 \text{ to } 0.216}$, the 0.99 interval for $p_N - p_A$

(b) No. The interval estimate contains the value zero.

Chapter Eleven

11-1 (a) $P(A) = P(B) = P(C) = P(D) = P(E) = 1/5$

(b) χ^2

(c) (i) Classical approach:

H_o: Equal preference

Polish	A	B	C	D	E	Total
Observed	27	17	15	22	19	100
Expected	20	20	20	20	20	100
$\frac{(O-E)^2}{E}$	49/20	9/20	25/20	4/20	1/20	88/20

$\chi^2 = 88/20 = 4.40$

Critical value: $\chi^2(4, 0.10) = 7.78$

Fail to reject H_o, the preferences of polish are not significantly different from equal proportions.

(ii) Prob-value approach:
The first several steps can be found in part (i).

$P = P(\chi^2 > 4.40$, with df $= 4)$, $\underline{0.10 < P < 0.50}$

11-4 (a) Classical approach:

H_o: This years distribution is same as previous

District	NE	SE	NC	SC	WC	Total
This yr.	120	128	43	66	143	500
Previous	100	140	40	60	160	500
$\frac{(O-E)^2}{E}$	4.000	1.029	0.2250	0.600	1.806	7.660

$\chi^2* = 7.66$

Critical value: $\chi^2(4, 0.05) = 9.50$

Fail to reject H_o, there is no significant difference between the distribution of this years sales and those of previous years.

(b) Prob-value approach:
The first several steps can be found in part (i).

$P = P(\chi^2 > 7.66$, with df $= 4)$, $\underline{0.10 < P < 0.50}$

11-6   The student leaders were able to give multiple answers and therefore the percentages reported total more than 100%. The multinomial experiment requires exactly one answer from each student.

11-8 H_o: The number of defective items is independent of the day of the week

$\chi^2(4, 0.05) = 9.50$

Expected values:

	Mon	Tues	Wed	Thur	Fri
Defect.	91	91	91	91	91
Do not	9	9	9	9	9

$\chi^{2*} = 0.396 + 0.011 + 0.176 + 0.176 + 0.011$
$+ 4.000 + 0.111 + 1.778 + 1.778 + 0.111$

$\chi^{2*} = 8.548$

Fail to reject H_o, there is not sufficient evidence to reject the hypothesis of independence.

11-11   (a) The information compares several distributions, a distribution for each region of New York State.

(b)   A test of homogeneity compares several distributions.

11-12   (a) Classical approach:

H_o: The distribution of reactions is the same for both groups.

$\chi^2(2, 0.10) = 4.61$

Expected values:

	Mild	Medium	Strong
Yes	144	120	36
No	96	80	24

$\chi^{2*} = 4.69 + 3.33 + 1.00 + 7.04$
$+ 5.00 + 1.50 = 22.56$

Reject H_a, yes there appears to be a relationship between neighborhood and reaction.

(b)   Prob-value approach:

The first several steps can be found in part (a).

$P = P(\chi^2 > 22.56,$ with df=2), $\underline{P < 0.005}$

11-15   (a) Classical approach:

H$_o$: Fear and Do Not Fear darkness are proportioned the same for each age group

$\chi^2(4, 0.01) = 13.3$

Expected values:

	Elem.	J. H.	S. H.	Coll.	Adult
Fear	70.8	70.8	70.8	70.8	70.8
Do not	129.2	129.2	129.2	129.2	129.2

$\chi^{2}* = 2.102 + 0.020 + 6.712 + 17.105 + 26.359$
$\quad\quad + 1.152 + 0.011 + 3.678 + 9.373 + 14.445$

$\chi^{2}* = 80.957$

Reject H$_o$, there is sufficient evidence to show that the age groups have different proportions which fear darkness.

(b)   Prob-value approach:

The first several steps can be found in part (a).

$P = P(\chi^2 > 80.96,$ with df=4), $\underline{P < 0.005}$

11-16   (a) Classical approach:

H$_o$: Distribution is 10, 20, 40, 20, 10 percent

Critical value: $\chi^2(4, 0.05) = 9.50$

	A	B	C	D	F
Observed	16	43	65	48	28
Expected	20	40	80	40	20

$\chi^{2}* = 0.800 + 0.225 + 2.812 + 1.600 + 3.200$
$\quad\quad = 8.637$

Fail to reject H$_o$, this grade distribution is not significantly different from the claimed distribution.

(b)   Prob-value approach:

The first several steps can be found in part (a).

$P = P(\chi^2 > 8.64,$ with df=4), $\underline{0.05 < P < 0.10}$

11-19 H_o: The weights are normally distributed with a mean of 160 and a standard deviation of 15 pounds.

That is: P(x<130)=0.0228,   P(130<x<145)=0.1359,
P(145<x<160)=0.3413, P(160<x<175)=0.3413,
P(175<x<190)=0.1359, P(x>190)=0.0228

Expected values:

	x<130	130<x<145	145<x<160	160<x<175	175<x<190	x>190
Obs	7	38	100	102	40	13
Exp	6.84	40.77	102.39	102.39	40.77	6.84
$\frac{(O-E)^2}{E}$	0.004	0.188	0.056	0.001	0.015	5.548

$\chi^{2}* = 5.812$

$P = P(\chi^{2}* > 5.812,$ with df $= 5) > 0.10$

Fail to reject H_o, there is not sufficient evidence to reject the null hypothesis that this data is normally distributed with $\mu = 169$ and $\sigma = 15$.

11-22 H_o: Independence

Expected values:

	20-35	36-50	over 50
Conservative	28.00	30.00	22.00
Moderate	73.50	78.75	57.75
Liberal	38.50	41.25	30.25

$\chi^{2}* = 2.286 + 3.333 + 0.182$
$+ 0.575 + 0.496 + 2.815$
$+ 0.058 + 6.402 + 7.192 = 23.339$

$P = P(\chi^2 > 23.339) < 0.005$

note: $(\chi^2(4,0.005)=16.8)$

Reject H_o, the sample evidence does contradict the claim of independence.

11-24 (a) Classical approach:

H_o: Proportion of popcorn that popped is the same for all brands.

$\chi^2(3,0.05) = 7.82$

Expected values:

	A	B	C	D
Popped	88	88	88	88
Not popped	12	12	12	12

$\chi^{2}* = 0.333 + 1.333 + 0.083 + 0.750$
$+ 0.045 + 0.182 + 0.011 + 0.102 = 2.839$

Fail to reject H_o, we are unable to reject the null hypothesis that the four brands pop equally well.

(b)   Prob-value approach:

   The first several steps can be found in part (a).

   $P = P(\chi^2 > 2.839, \text{ with } df=3)$, $\underline{0.10 < P < 0.50}$

11-27   (a) Classical approach:

   H_0: Rate of absenteeism is the same for all groups.

   $\chi^2(3, 0.01) = 11.4$

   Expected values:

Days	Married Male	Single Male	Married Female	Single Female
Absent	200	70	80	150
Worked	9400	3290	3760	7050

   $\chi^2* = 2.0000 + 22.8571 + 0.3125 + 1.5000$
   $+ 0.0425 + 0.4863 + 0.0066 + 0.0319 = 27.2369$

   Reject H_0, there is a different rate of absenteeism among the categories of employees.

(b)   Prob-value approach:

   The first several steps can be found in part (a).

   $P = P(\chi^2 > 27.2, \text{ with } df=3)$, $\underline{P < 0.005}$

11-29   (a)

$$z = \frac{\frac{75}{100} - \frac{70}{100}}{\sqrt{(\frac{75+70}{100+100})(1-\frac{145}{200})(\frac{1}{100}+\frac{1}{100})}}$$

$$= \frac{0.750 - 0.700}{\sqrt{(0.725)(0.275)(0.01+0.01)}}$$

   $z* = \underline{0.7918}$

(b)   Expected values:

	Yes	No
Group 1	72.5	27.5
Group 2	72.5	27.5

   $\chi^2* = 0.0862 + 0.2272 + 0.0862 + 0.2272 = \underline{0.6268}$

(c) $\chi^2 = 0.6268$

   $(t*)^2 = (0.7918)^2 = 0.6269$

Chapter Twelve

12-1 (a) The mean levels of the test factor are not all the same.

(b) The mean levels of the test factor are all the same.

(c) The mean levels of the test factor are all the same.

12-2 df(factor) appears first in the critic number notation since ms(factor) is the numerator for the calculated value of the test statistic F.

12-3 (a) 0   (b) 3   (c) 16   (d) 60   (e) 1,232

12-5 (a) H_o: The mean values for each of the levels of the tested factor are all equal.

H_a: The mean values for each of the levels of the tested factor are not all equal. That is, at least one is different in value from the rest.

(b) We would conclude that the alternative hypothesis is correct.

(c) We would conclude that the evidence found was not sufficient to contradict the null hypothesis.

(d) The decision is made using an F test by comparing the calculated F* with the critical value of F obtained from table 8.

12-9 H_o: The mean values for workers are all equal.

$\alpha = 0.05$

Sample information:
$n = 15$, $C_1 = 46$, $C_2 = 58$, $C_3 = 57$, $T = 161$
$\Sigma x^2 = 1771$

Source	SS	df	MS
Workers	17.73	2	8.87
Error	25.20	12	2.10
Total	42.93	14	

} F* = 4.22

Critical value: $F(2, 12, 0.05) = 3.89$

Reject H_o: there is a significant difference between the workers with regards to mean amount of work produced.

12-11 H_0: The mean typing speeds on both types of typewriters are equal.

$\alpha = 0.05$

Sample information:
$n = 12$, $C_1 = 357$, $C_2 = 299$, $T = 656$
$\Sigma x^2 = 36948$

Source	SS	df	MS	
Typewriter	280.333	1	280.333	} $F^* = 3.477$
Error	806.333	10	80.633	
Total	1086.666	11		

Critical value: $F(1,10,0.05) = 4.96$

Fail to reject H_0: there is no significant difference between the typewriters with regards to mean typing speed.

12-14 H_0: There is no difference in the yield due to the three concentrations.

$\alpha = 0.05$

Sample information:
$n = 15$, $C_1 = 217$, $C_2 = 263$, $C_3 = 180$,
$T = 660$, $\Sigma x^2 = 29846$

Source	SS	df	MS	
Concent.	691.60	2	345.80	} $F^* = 36.27$
Error	114.40	12	9.53	
Total	806.00	14		

Critical value: $F(2,12,0.05) = 3.89$

Reject H_0: there is a significant difference in the yield resulting from the use of the three concentrations.

12-17 H_0: The mean stopping distance is not affected by the brand of tire.

$\alpha = 0.05$

Sample information:
$n = 23$, $C_1 = 217$, $C_2 = 194$, $C_3 = 216$,
$C_4 = 245$, $T = 872$, $\Sigma x^2 = 33282$

Source	SS	df	MS	
Brand	95.359	3	31.7865	} $F^* = 4.78$
Error	126.467	19	6.6562	
Total	221.826	22		

Critical value: $F(3,19,0.05) = 3.13$

Reject H_0: the brand of tire used does have a significant effect on mean stopping distance.

12-20 H_0: The mean amounts dispensed by the machines are all equal.

$\alpha = 0.01$

Sample information:
$n = 18$, $C_1 = 16.5$, $C_2 = 20.6$, $C_3 = 16.9$,
$C_4 = 19.1$, $C_5 = 21.8$, $T = 94.9$, $\Sigma x^2 = 523.49$

Source	SS	df	MS	
Machine	20.998	4	5.2495	} F* = 31.6
Error	2.158	13	0.166	
Total	23.156	17		

Critical value: $F(4,13,0.01) = 5.21$

Reject H_0: there is a significant difference between the machines with regards to mean amount of soft drink dispensed.

Chapter Thirteen

13-1   Refer to the definition and development of the measures of central tendency, section 2-3, and the measures of dispersion, section 2-4.

13-3   (a)

[Scatter plot: y-axis 0 to 6, x-axis 1 to 9, showing data points]

(b)

x	y	$x-\bar{x}$	$y-\bar{y}$	$(x-\bar{x})(y-\bar{y})$
1	1	-4	-2.5	10
1	2	-4	-1.5	6
3	2	-2	-1.5	3
3	3	-2	-0.5	1
5	3	0	-0.5	0
5	4	0	0.5	0
7	4	2	0.5	1
7	5	2	1.5	3
9	5	4	1.5	6
9	6	4	2.5	10
50	35	0	0.0	40

$\bar{x} = 50/10 = 5.0$, $\bar{y} = 35/10 = 3.5$

$\text{Covar}(x,y) = 40/9 = \underline{4.44}$

Summary of sample data: $n = 10$, $\Sigma x = 50$, $\Sigma y = 35$, $\Sigma x^2 = 330$, $\Sigma xy = 215$, $\Sigma y^2 = 145$

(c) $s_x = \sqrt{\dfrac{330 - (50)^2/10}{9}} = \sqrt{8.889} = \underline{2.981}$

$s_y = \sqrt{\dfrac{145 - (35)^2/10}{9}} = \sqrt{2.50} = \underline{1.581}$

(d) $r = \dfrac{4.44}{(2.981)(1.581)} = \underline{0.943}$

(e) $SS(x) = 330 - (50)^2/10 = 80$
$SS(y) = 145 - (35)^2/10 = 22.5$
$SS(xy) = 215 - (50)(35)/10 = 40$

$r = 40/\sqrt{80}\sqrt{22.5} = 0.9428 = \underline{0.943}$

13-5 (a)

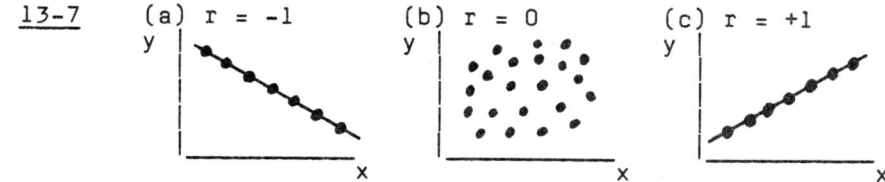

(b)

x	y	x-x̄	y-ȳ	(x-x̄)(y-ȳ)
0	6	-3.5	2.5	-8.75
1	6	-2.5	2.5	-6.25
1	7	-2.5	3.5	-8.75
2	4	-1.5	0.5	-0.75
3	5	-0.5	1.5	-0.75
4	2	0.5	-1.5	-0.75
5	3	1.5	-0.5	-0.75
6	0	2.5	-3.5	-8.75
6	1	2.5	-2.5	-6.25
7	1	3.5	-2.5	-8.75
35	35	0.0	0.0	-50.50

$\bar{x} = 35/10 = 3.5$, $\bar{y} = 35/10 = 3.5$

$\text{Covar}(x,y) = -50.5/9 = \underline{-5.61}$

(c) $s_x = \sqrt{54.5/9} = \sqrt{6.1} = \underline{2.46}$

$s_y = \sqrt{54.5/9} = \sqrt{6.1} = \underline{2.46}$

(d) $r = \dfrac{-5.61}{(2.46)(2.46)} = \underline{-0.93}$

(e) $SS(x) = 170 - (35)^2/10 = 54.5$
$SS(y) = 177 - (35)^2/10 = 54.5$
$SS(xy) = 72 - (35)(35)/10 = -50.5$

$r = -50.5/\sqrt{54.5}\sqrt{54.5} = \underline{-0.93}$

13-7 (a) $r = -1$ (b) $r = 0$ (c) $r = +1$

13-8 $H_0: \rho = 0.0$

$H_a: \rho \neq 0.0$

$\alpha = 0.10$

$r^* = 0.43$

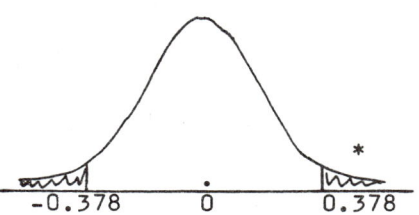

Reject H_0, there is sufficient reason to reject the null hypothesis that $\rho = 0$.

13-9 $H_0: \rho = 0.0$

$H_a: \rho < 0.0$

$\alpha = 0.01$

$r^* = -0.50$

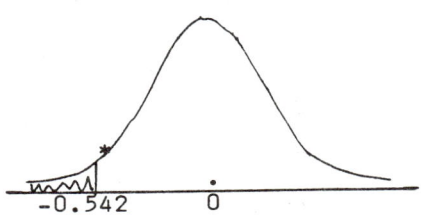

Fail to reject H_0, there is not sufficient reason to reject the null hypothesis.

13-12   (a) -0.55 to 0.75   (b) -0.52 to -0.22
        (c) 0.32 to 0.82   (d) -0.67 to 0.33

13-14
(a)
Annual Price and Expenditures for a New Car

Detroit Hangs Tough

(b)   0.9   Yes, these variables seem to be linearly correlated. The two graphs shown in Case Study 13.2 have almost identical patterns and the scatter diagram drawn in (a) shows the ordered pairs following a strong upward pattern.

13-18 Summary of sample data:
$n = 20$, $\quad \Sigma x = 302$, $\quad \Sigma y = 455$,
$\Sigma x^2 = 5{,}148$, $\quad \Sigma xy = 7{,}120$, $\quad \Sigma y^2 = 10{,}837$

$SS(x) = 5{,}148 - (302^2/20) = 587.8$
$SS(xy) = 7{,}120 - \{(302)(455)/20\} = 249.5$

Using formula (3-7)
$b_1 = 249.5/587.8 = 0.4245$

Using formula (3-6)
$b_0 = \frac{1}{20}\{455 - (0.4245)(302)\} = 16.3406$

$\hat{y} = 16.34 + 0.4245x$

$s_e^2 = \dfrac{10837 - (16.3406)(455) - (0.4245)(7120)}{18}$

$s_e^2 = 379.587/18 = \underline{21.088}$

13-21 (a) Hours Studied vs. Exam Grade

Exam Grade

Hours Studied

Summary of sample data: $n = 15$, $\Sigma x = 81$, $\Sigma y = 110$,
$\Sigma x^2 = 487$, $\Sigma xy = 625$, $\Sigma y^2 = 842$

(b) Using formula (2-9)
$SS(x) = 487 - (81^2/15) = 49.6$
Using formula (3-4)
$SS(xy) = 625 - \{(81)(110)/15\} = 31$

Using formula (3-7)
$b_1 = 31/49.6 = 0.625$
Using formula (3-6)
$b_0 = \frac{1}{15}\{110 - (0.625)(81)\} = 3.96$

$\hat{y} = 3.96 + 0.625x$

(c) If $x = 2$, then $\hat{y} = 3.96 + 0.625(2) = \underline{5.22}$
If $x = 3$, then $\hat{y} = 3.96 + 0.625(3) = \underline{5.85}$
If $x = 4$, then $\hat{y} = 3.96 + 0.625(4) = \underline{6.48}$
If $x = 5$, then $\hat{y} = 3.96 + 0.625(5) = \underline{7.11}$
If $x = 6$, then $\hat{y} = 3.96 + 0.625(6) = \underline{7.74}$
If $x = 7$, then $\hat{y} = 3.96 + 0.625(7) = \underline{8.37}$
If $x = 8$, then $\hat{y} = 3.96 + 0.625(8) = \underline{9.00}$

(d)

$e = y - \hat{y}$					
x	3	3	6	6	6
y	5	7	6	9	8
$\hat{y}$	5.85	5.85	7.74	7.74	7.74
e	-0.85	1.15	-1.74	1.26	0.26

(e) $s_e^2 = \dfrac{842-(3.96)(110)-(0.625)(625)}{13} = 12.65/13 = \underline{0.973}$

13-23 (a)

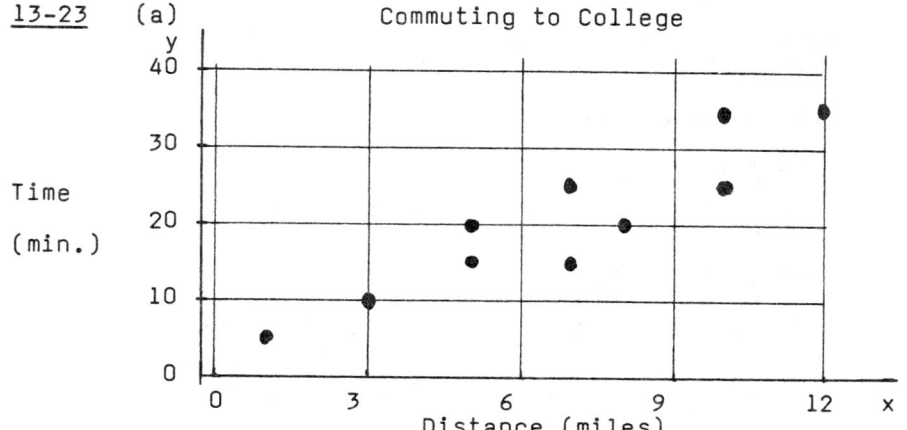

Commuting to College

Summary of sample data: $n = 10$, $\Sigma x = 68$, $\Sigma y = 205$, $\Sigma x^2 = 566$, $\Sigma xy = 1670$, $\Sigma y^2 = 5075$

(b) Using formula (2-9)
$SS(x) = 566 - (68^2/10) = 103.6$
Using formula (3-4)
$SS(xy) = 1670 - \{(68)(205)/10\} = 276.0$
Using formula (3-7)
$b_1 = 276.0/103.6 = 2.664$
Using formula (3-6)
$b_0 = \frac{1}{10}\{205 - (2.664)(68)\} = 2.38$

$\underline{\hat{y} = 2.38 + 2.664x}$

(c) $H_o: \beta_1 = 0.0$

$H_a: \beta_1 > 0.0$

$\alpha = 0.05$

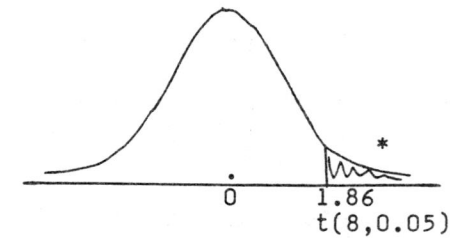

1.86
$t(8, 0.05)$

continued

$$s_e^2 = \frac{5075-(2.38)(205)-(2.664)(1670)}{8}$$

$s_e^2 = 137.192/8 = 17.149$

$s_{b_1}^2 = 17.149/103.6 = 0.16553$

$s_{b_1} = \sqrt{0.16553} = 0.40685$

$t^* = \frac{2.66 - 0}{0.40685} = 6.54$

Reject H_o, there is sufficient reason to conclude that β_1 is greater than zero.

(d) $2.66 \pm (2.60)(0.40685)$

2.66 ± 1.18

1.48 to 3.84, the 0.98 interval for β_1

13-25   Summary of sample data:
n = 10 $\Sigma x = 16.25$, $\Sigma y = 152$,
$\Sigma x^2 = 31.5625$, $\Sigma xy = 275$, $\Sigma y^2 = 2504$

$SS(x) = 31.5625 - (16.25^2/10) = 5.15625$
$SS(xy) = 275 - \{(16.25)(152)/10\} = 28.0$

Using formula (3-7)
$b_1 = 28.0/5.15625 = 5.4303$

Using formula (3-6)
$b_o = \frac{1}{10}\{152 - (5.4303)(16.25)\} = 6.3758$

$\hat{y} = 6.3758 + 5.4303x$

$$s_e^2 = \frac{2504 - (6.3758)(152) - (5.4303)(275)}{8}$$

$s_e^2 = 5.19324$

$s_e = \sqrt{5.19324} = 2.279$

$\hat{y}$ at $x = 2.00$ is $\hat{y} = 6.3758 + 5.4303(2.00) = 17.24$

$t(8, 0.025) = 2.31$

$\bar{x} = 1865/25 = 74.6$

(a) $17.24 \pm (2.31)(2.278)\sqrt{\frac{1}{10} + \frac{(2-1.625)^2}{5.15625}}$
$17.24 \pm (2.31)(2.278)\sqrt{0.1272727}$
17.24 ± 1.88

15.4 to 19.1, the 0.95 interval for $\mu_{y|x=2}$

(b) $17.24 \pm (2.31)(2.278)\sqrt{1 + 0.1272727}$
17.24 ± 5.59

11.6 to 22.8, the 0.95 interval for $y_{x=2}$

13-27   The standard deviation for $\bar{x}$'s is much smaller than for individual x's (Central Limit Theorem). Thus the confidence interval will be narrower in accordance to this.

13-28   Yes. $s\sqrt{1/n} = s/\sqrt{n}$ and that is the estimate for the standard error of the mean.

13-29   (a) Always

(b)   Never. r = 0.99 only indicates a strong correlation. It never indicates cause-effect.

(c)   Sometimes. An r value greater than zero indicates that as x increase, y tends to increase. However, there may be a few high x-values with low y-values.

(d)   Sometimes. The two coefficients measure two completely different concepts. Their signs are unrelated.

(e)   Always

13-32   (a)

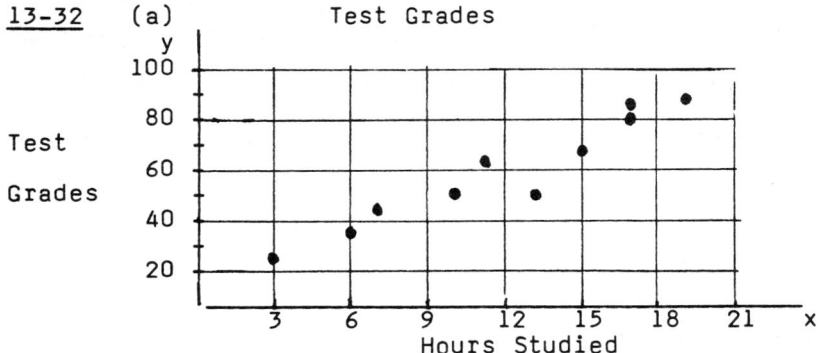

Estimate r at 0.9

Summary of sample data: n = 10, Σx = 118, Σy = 591, Σx^2 = 1648, Σxy = 7956, Σy^2 = 39013

(b)   SS(x) = 1648 - (118)²/10 = 255.6
        SS(y) = 39013 - (591)²/10 = 4084.9
        SS(xy) = 7956 - (118)(591)/10 = 982.2

        r = 982.2/√255.6√982.2 = <u>0.961</u>

(c) $H_0: \rho = 0.0$

        $H_a: \rho > 0.0$

        $\alpha = 0.01$

        $r^* = 0.961$

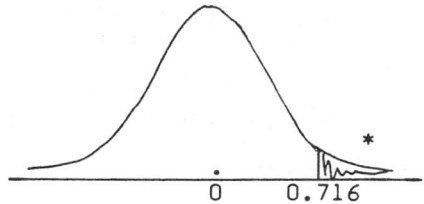

Reject H_0, there is sufficient reason to conclude that ρ is positive.

(d)   **0.85 to 0.98**, the 0.95 interval for ρ

13-35   (a) Summary of sample data:
$n = 10$, $\Sigma x = 21$, $\Sigma y = 763$,
$\Sigma x^2 = 51$, $\Sigma xy = 1538$, $\Sigma y^2 = 58937$

$SS(x) = 51 - (21)^2/10 = 6.9$
$SS(y) = 58937 - (763)^2/10 = 720.1$
$SS(xy) = 1538 - (21)(763)/10 = -64.3$

$r = -64.3/\sqrt{6.9}\sqrt{720.1} = \underline{-0.912}$

(b)   Summary of sample data:
$n = 10$, $\Sigma x = 19$, $\Sigma y = 763$,
$\Sigma x^2 = 43$, $\Sigma xy = 1514$, $\Sigma y^2 = 58937$

$SS(x) = 43 - (19)^2/10 = 6.9$
$SS(y) = 58937 - (763)^2/10 = 720.1$
$SS(xy) = 1514 - (19)(763)/10 = 64.3$

$r = 64.3/\sqrt{6.9}\sqrt{720.1} = \underline{0.912}$

(c)   They are numerically equal, but of opposite sign.

13-38   (a) Prefinal Average vs. Final Exam Score

[Scatter plot: Final Exam Score (y-axis, 45 to 75) vs. Prefinal Average (x-axis, 55 to 95), with fitted line labeled (b), point labeled (e)]

(b)   See line on graph in (a), $\hat{y} = 25 + 0.5x$

-95-

(c) Estimate r at 0.75

Summary of sample data: n = 25, $\sum x$ = 1865, $\sum y$ = 1563
$\sum x^2$ = 142771, $\sum xy$ = 118392, $\sum y^2$ = 99019

(d) Using formula (2-9)
SS(x) = 142771 - (1865^2/25) = 3642
Using formula (3-4)
SS(xy) = 118392 - {(1865)(1563)/25} = 1792.2
Using formula (3-7)
b_1 = 1792.2/3642 = 0.492
Using formula (3-6)
b_0 = $\frac{1}{25}${1563 - (0.492)(1865)} = 25.82

$\hat{y}$ = 25.82 + 0.492x

(e) See line on graph in (a)

(f) SS(x) = 3642, SS(xy) = 1792.2 (see part d)
SS(y) = 99019 - (1563^2/25) = 1300.24

r = 1792.2/$\sqrt{3642.0}\sqrt{1300.24}$ = 0.8236 = 0.82

(g) H_0: ρ = 0.0

H_a: $\rho \ne$ 0.0

α = 0.10

r* = 0.824

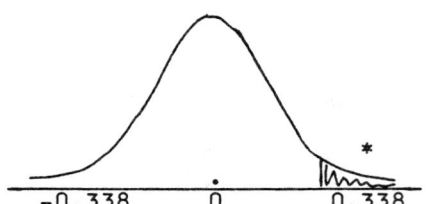

-0.338   0   0.338

Reject H_0, there is sufficient reason to conclude that there is a correlation.

(h) 0.65 to 0.92, the 0.95 interval for ρ (from table 10)

(i) s_e^2 = $\frac{99019-(25.82)(1563)-(0.492)(118392)}{25 - 2}$

s_e^2 = 417.42/23 = 18.149

s_e = $\sqrt{18.149}$ = 4.26

(j) $s_{b_1}^2$ = 18.149/3642 = 0.004983

s_{b_1} = $\sqrt{0.004983}$ = 0.0706

0.492 $\pm$ (2.07)(0.0706)

0.492 $\pm$ 0.146

0.346 to 0.638, the 0.95 interval for β_1

(k) $H_0: \beta_1 = 0.0$

$H_a: \beta_1 > 0.0$

$\alpha = 0.05$

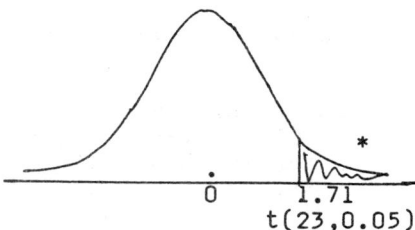

$t^* = \dfrac{0.492 - 0}{0.0706} = 6.97$

Reject H_0, there is sufficient reason to conclude that β_1 is significantly greater than zero.

(l) At $x = 85$, $\hat{y} = 25.81 + 0.492(85) = 67.63$

$\bar{x} = 1865/25 = 74.6$

$67.63 \pm (2.07)(4.26)\sqrt{\dfrac{1}{25} + \dfrac{(85-74.6)^2}{3642}}$

$67.63 \pm (2.07)(4.26)\sqrt{0.06998}$
67.63 ± 2.33

65.30 to 69.96, the 0.95 interval for $\mu_{y|x=85}$

(m) At $x = 78$, $y = 25.81 + 0.492(78) = 64.19$

$64.19 \pm (2.07)(4.26)\sqrt{1 + \dfrac{1}{25} + \dfrac{(78-74.6)^2}{3642}}$

$64.19 \pm (2.07)(4.26)\sqrt{1.04317}$
64.19 ± 9.01

55.18 to 73.20, the 0.95 interval for John, $y_{x=85}$

<u>13-41</u> Summary of data:
$n = 3$, $\Sigma x = 3$, $\Sigma y = 6$, $\Sigma x^2 = 5$, $\Sigma xy = 7$, $\Sigma y^2 = 14$
$SS(x) = 5 - 3^2/3 = 2$, $SS(xy) = 7 - ((3)(6)/3) = 1$
$b_1 = 1/2 = 0.5$, $b = \dfrac{1}{3}(6 - (0.5)(3)) = 1.5$

$\hat{y} = 1.5 + 0.5x$ $\bar{y} = y/n = 6/3 = 2$

x	y	$\hat{y}$	$y-\bar{y}$	$(y-\bar{y})^2$	$y-\hat{y}$	$(y-\hat{y})^2$	$\hat{y}-\bar{y}$	$(\hat{y}-\bar{y})^2$
0	1	1.5	1.0	1.00	-0.5	0.25	-0.5	0.25
1	3	2	1.0	1.00	1.0	1.00	0.0	0.00
2	2	2.5	0.0	0.00	-0.5	0.25	0.5	0.25
sum				2.00		1.50		0.50

Therefore, $\Sigma(y-\bar{y})^2 = \Sigma(y-\hat{y})^2 + \Sigma(\hat{y}-\bar{y})^2$

Chapter Fourteen

14-1 (a) $H_0: M = 48$

(b) $H_0: P(+) = 0.5$ (+ stands for above 48)

(c) $H_0: M = 48$

$H_a: M \neq 48$

$n(+) = 3$, $n(-) = 16$, $n(0) = 1$, $n = 19$ and $x = 3$

$\alpha = 0.05$

```
         *
/////////|_____
0 1 2 3 4                   x
```

Reject H_0, the median teperature appears to be different from 48.

14-3 H_0: Equal preference for whole wheat and white flour crust for pizza

H_a: Whole wheat flour crust is preferred

preferences:
(+) = whole wheat preferred, (−) = white flour preferred, 0 = no preference
$n(+) = 65$, $n(-) = 53$, $n(0) = 15$, $n = 118$, $x = 65$

$x' = 64.5$

$z = \dfrac{64.5 - (118/2)}{(1/2)\sqrt{118}}$

$z^* = 1.01$

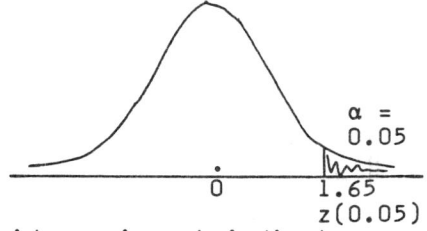

$\alpha = 0.05$

Fail to reject H_0, the evidence is not indicate a preference for whole wheat crust pizza.

14-5 For $n = 20$ and $1 - \alpha = 0.95$, the critical value from table 11 is $k = 5$.

$x_{k+1} = x_6 = 39$ and $x_{n-k} = x_{15} = 47$

<u>39 to 47</u>, the 0.95 interval for median M

14-8 H_0: No difference in the reaction times.

H_a: There is a difference.

$\alpha = 0.05$ (two-tailed), $n_1 = 9$, $n_2 = 6$,
critical value = 10

```
                                          ////////|
                                          0 ... 10
```

ranked data	4.9	5.0	5.1	5.5	5.5	5.7	5.9	6.0
rank	1	2	3	4.5	4.5	6	7	9
source	1	1	1	1	1	1	2	1

	6.0	6.0	6.2	6.5	7.0	7.2	7.2
	9	9	11	12	13	14.5	14.5
	2	1	1	2	2	2	2

$R_1 = 50$, $R_2 = 70$

$U_1 = (9)(6) + \frac{(6)(7)}{2} - 70 = 5$

$U_2 = (9)(6) + \frac{(9)(10)}{2} - 50 = 49$ } $U = 5$

The observed value is less than the critical
value, therefore, reject H_0, and conclude
that there is a significant difference in the
reaction times

14-11 H_0: No difference between the average systolic
        blood pressure of men of age 40 and men age 25.

H_a: There is a difference.

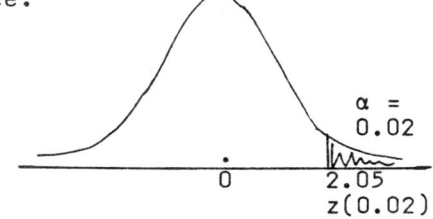

$\alpha = 0.02$

$z(0.02)$

$R_{25} = 540$, $R_{40} = 945$

$U_{25} = (24)(30) + \frac{(30)(31)}{2} - 945 = 240$

$U_{40} = (24)(30) + \frac{(24)(25)}{2} - 540 = 480$ } $U = 480$

$\mu_U = (24)(30)/2 = 360$

$\sigma_U = \sqrt{\frac{(24)(30)(24+30+1)}{12}} = 57.446$

$z^* = \frac{480 - 360}{57.446} = 2.09$

Reject H_0, there is a significant difference
in the systolic blood pressures of the two age groups.

14-13 H_0: The hiring sequence is random.

H_a: The hiring sequence is not of random order.

$\alpha = 0.05$, $n(M) = 15$, $n(F) = 5$

Critical values of V are 4 and 12

Observed V = 9

Fail to reject H_0, we would not be correct in concluding that this sequence is not random.

14-17 H_0: Randomness in number of absences (about median)

H_a: Lack of randomness

(1) $\alpha = 0.05$,   n(above) = 13,   n(below) = 13

critical values: 8 and 20

median = 10.5,   (a = above)

b a b b a a a a a a a b b b a a b b b b b b a a b

V = 9

Fail to reject H_0, the number of absences do not show a significant lack of randomness.

(2) $\alpha = 0.05$

$$\mu_V = \frac{2(13)(13)}{13+13} + 1 = 14$$

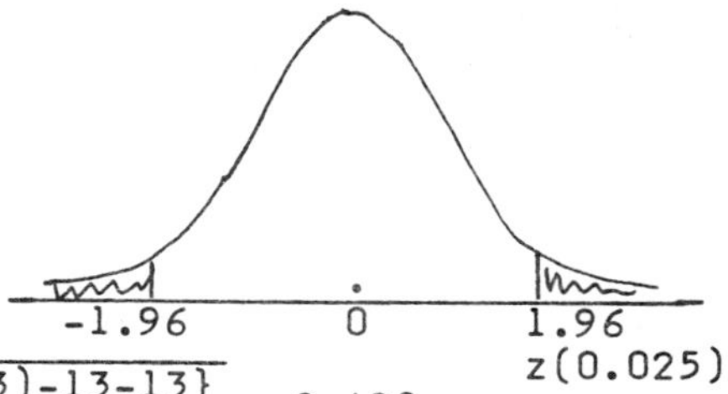

$$\sigma_V = \sqrt{\frac{2(13)(13)\{2(13)(13)-13-13\}}{(13+13)^2(13+13-1)}} = 2.498$$

$$z^* = \frac{9-14}{2.498} = -2.00$$

Reject H_0: there is evidence of lack of randomness.

14-19   (a)

x	rank of x	y	rank of y	d	d²
10	7	30	7	0	0.0
12	8	60	9.5	-1.5	2.25
15	11	50	8	3	9.0
5	3	12	4	-1	1.0
7	5	10	2	3	9.0
5	3	25	5.5	-2.5	6.25
5	3	10	2.5	1	1.0
15	11	60	9.5	1.5	2.25
8	6	25	5.5	0.5	0.25
1	1	10	2	-1	1.0
13	9	75	11	-2	4.0
15	11	95	12	1	1.0
				sum =	37.0

$$r_s = 1 - \frac{6(37.0)}{12(144-1)} = \underline{0.87}$$

(b) $H_0: \rho_s = 0$

$H_a: \rho_s > 0$

$\alpha = 0.05$

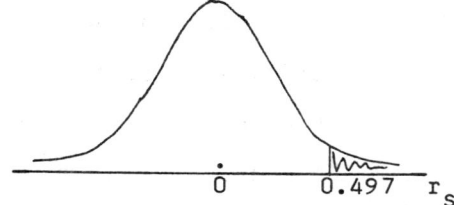

$r_s^* = 0.87$ is greater than 0.497, therefore,

Reject H_0, and conclude that there is a positive correlation.

14-22   (a)

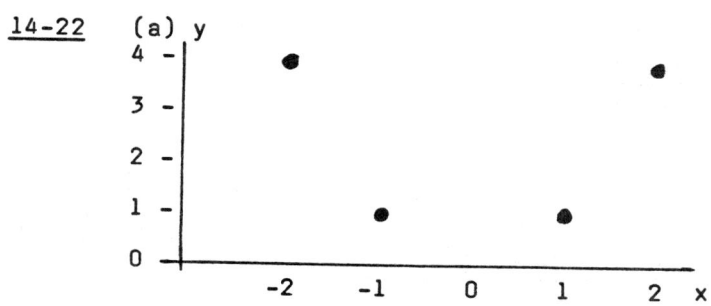

(b)

x	y	rank of x	rank of y	d	d²
-2	4	1	3.5	-2.5	6.25
-1	1	2	1.5	0.5	0.25
1	1	3	1.5	1.5	2.25
2	4	4	3.5	0.5	0.25
					9.00

$$r_s = 1 - \frac{6(9.0)}{4(16-1)} = \underline{0.10}$$

(c)

x	y	x^2	xy	y^2
-2	4	4	-8	16
-1	1	1	-1	1
1	1	1	1	1
2	4	4	8	16
0	10	10	0	34

$$r = \frac{0 - \{(0)(10)/4\}}{\sqrt{10-0^2/4}\sqrt{34-10^2/4}} = \underline{0.00}$$

(d) The two results are not identical, but both coefficients are near or at zero. The rank correlation measures the correlation between the rank numbers, while the Pearson coefficient uses the numerical values.

14-24 (a) H_o: No difference between average times (no faster)

H_a: Average time on B is less than on A (B is faster)

$\alpha = 0.05$ (one-tailed)

$n(+) = 2$, $n(-) = 8$, $n = 10$ and $x = 2$

```
//////|           *
0     1
```

(b) Fail to reject H_o, the evidence is not sufficient to justify the claim that track B is faster.

14-27 H_o: No difference

H_a: Top ripens first

'+' = if top ripens first, '-' = if bottom ripens first, '0' = if tied

$n(+) = 14$, $n(-) = 4$, $n(0) = 2$, Observed $x = 4$

$\alpha = 0.05$ (one-tailed), $n = 18$

Critical value is 5

Reject H_o, there is sufficient reason to conclude that the apples in the top of the tree start to ripen before those on the bottom half of the tree.

14-29 H_0: No effect

H_a: Computer assisted instruction produced higher achievement

$\alpha = 0.05$

Ranked data	55	59	60	65	70	72	75	77	77
rank	1	2	3	4	5	6	7	8.5	8.5
scource	1	2	1	2	1	2	1	1	2

79	83	85	88	89	90	90	90	92	92	92
10	11	12	13	14	16	16	16	19	19	19
1	1	2	1	1	2	2	1	1	2	2

$R_1 = 107.5$, $R_2 = 102.5$

$U_1 = (10)(10) + \frac{(10)(11)}{2} - 102.5 = 52.5$

$U_2 = (10)(10) + \frac{(10)(11)}{2} - 107.5 = 47.5$

$U = 47.5$

The observed value is greater than the critical value of 27, therefore, fail to reject H_0, and cannot conclude that the computer assisted instruction produced higher achievement scores.

14-31 H_0: Random order

H_a: Lack of randomness

$\alpha = 0.05$, $n(n) = 20$, $n(d) = 4$

Critical values of V are 4 and 10

Observed V = 9

Fail to reject H_0, the sample results do not show any lack of randomness.

14-34 H_0: $\rho_s = 0$ (not correlated)

H_a: $\rho_s \neq 0$ (correlated)

$\alpha = 0.05$

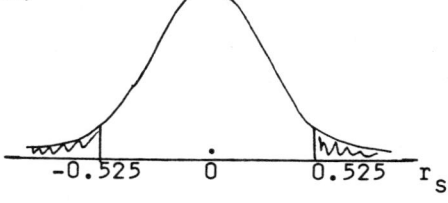

d: -1 3.5 1 -5 1 2.5 0.5 -1.5 4 2 1.5 0.5 1 2 -12

$n = 15$, $\Sigma d^2 = 220.50$

$r_s = 1 - \frac{6(220.5)}{15(15^2-1)} = 0.39$

Fail to reject H_0, there is no significant correlation shown between the two sets of scores.